$15.—

440

D150806.3

IRON FIST

Other Books by Bryan Perrett

IRON FIST

Classic Armoured Warfare Case Studies
BRYAN PERRETT

'Before a war military science seems a real science, like astronomy;
after a war it seems more like astrology.'

— Rebecca West, novelist, 1892-1983

'It is not, of course, that the tank offers protection to those who
fight in it. A trench or a hole in the ground will do the same. The
tank is essentially a mobile weapon of *offence*.'

— Major Clough Williams-Ellis, *The Tank Corps*, 1919

'The commander must be at constant pains to keep his troops
abreast of all the latest tactical experience and developments and
must insist on their practical application. He must see to it that his
subordinates are trained in accordance with the latest requirements.'

— Field Marshal Erwin Rommel, *The Rommel Papers*, 1953

Brockhampton Press

Arms and Armour Press
A Cassell Imprint
Wellington House, 125 Strand, London WC2R 0BB

This edition published in 1999 by Brockhampton Press,
a member of Hodder Headline PLC Group

ISBN 1 86019 954 2

British Library Cataloguing-in-Publication Data:
a catalogue record for this book is available from the British Library

Designed and edited by DAG Publications Ltd.
Designed by David Gibbons; layout by Anthony A. Evans;
edited by Gerald Napier.
Printed and bound in Great Britain by
Creative Print and Design (Wales), Ebbw Vale

Contents

Acknowledgements

should like to express my sincere appreciation and thanks to the following for their generous assistance: Major J. Charlton-Jones, The Queen's Own Yeomanry; David Fletcher of the Tank Museum, Bovington; Mr Nicholas Forder, Derby City Council Museum and Art Gallery; Colonel Paul Gaujac, Chef du Service Historique de l'Armée de Terre, Vincennes; Hauptmann Gercke, Panzerbataillon 84, German Army; Lieutenant Colonel D. J. K. German and Major J. J. Eadie, Museum of the Staffordshire Yeomanry, Stafford; Mr Gareth S. Gill, Curator, 1st The Queen's Dragoon Guards Regimental Museum, Cardiff; Colonel M. B. Haycock, The Warwickshire Yeomanry Museum, Warwick; Major A. W. Kersting, The Household Cavalry Museum, Windsor; Oberstleutnant Joh. Kindler, Militargeschichtliches Forschungsamt, German Army; Lieutenant Colonel R. B. Merton, Lieutenant Colonel A. R. D. Shirreff, Major D. A. J. Williams and Captain S. T. W. Bridge, The King's Royal Hussars; Oberstleutnant Paprotka, Wehrgeschichtliches Museum, Rastatt; Mr John M. Purdy and SFC John T. Broom of the Patton Museum of Cavalry and Armor, Fort Knox, Kentucky; and to Pan Macmillan and Penguin USA for permission to use the passage quoted from Alistair Horne's *To Lose a Battle* on page 73.

Introduction

A s with every area of human activity, the evolution of armoured warfare over the past 90 years has been a process of continual learning and, within this, one of the most important elements has been what might be called in jargon the 'man-machine interface'. To say that the fighting vehicle and its crew are complementary is, on the one hand, an absurd over-statement of the obvious yet, on the other, the relationship does extend well beyond the bounds of simple ergonomics. From the very beginning a dialogue has existed between the user and the designer. For example, during the first weeks of World War I the crews of the Belgian and Royal Naval Air Service touring cars requested the protection of armour plate; the latter also requested fully-rotating turrets in which to mount their machine-guns, a most important development although the idea was far from new. Simultaneously, there developed a second dialogue, that between the user and those under whose orders he operated. This considered such factors as the vehicle's limitations, the terrain to be fought over, and the threat posed by the enemy's probable response. Together, these influences have been reflected in the design of successive generations of fighting vehicles ever since. Until comparatively recent times, the story has been one of reaction to events, with users often having to wait months or years for the improvements or new vehicles which their experience had indicated were essential; and, of course, when these did reach the front, history had often moved forward a step or two. Thus, the crews who took the Tank Mark I into action in September 1916 would have given their eye teeth for the modifications incorporated in the Tank Mark V of 1918; the tanks of 1945 looked nothing like those of 1939 and had been designed with very different requirements in mind; and the veteran of World War II, entering a modern tank, finds that it incorporates systems that in his youth were regarded as belonging to the realms of science fiction.

My intentions in writing this book, therefore, have been to show how the vehicles themselves and the tactics employed were developed in the light of the crews' experience, and in so doing give the reader some impression of what the tactical tank battle looked like from the inside as it evolved over the years. The method I have employed has been to select one or more episodes from each of the critical periods of history and place the emphasis on the human dimension of these. The higher thought processes of armoured warfare, the strategy and the more technical aspects of the subject have all received extensive coverage elsewhere and have only been included insofar as they are relevant to each episode.

I have devoted what I believe to be the appropriate amount of space to World War I for the very good reason that while the tank was originally conceived as a tool of trench warfare its use stimulated ideas among far-sighted officers who recognised its full potential. Likewise, each action fought during those early days added a paragraph to the hitherto blank page of experience.

Even so, on the outbreak of World War II only two armies, the British and German, had evolved clear-cut ideas as to how armoured formations could be used as a striking force within their own right. The former, thanks to years of political neglect, was temporarily hamstrung; the latter was not and quickly defeated Poland and the Western Allies in turn. Despite this, the German High Command was acutely aware of the risks involved and extremely sensitive to any check it received. Thus, if the tank battles in the Gembloux Gap and at Flavion had resulted in French victories, the 1940 campaign in the west would almost certainly have followed a different course.

In 1941 Hitler rashly decided to attack the Soviet Union. His armoured troops had become masters of the tactical battle yet, to their horror, they discovered that the latest generation of Russian tanks were superior to their own. From such encounters as that at Raseiniai and the untidy tank mêlée at Brody-Dubno grew a spiral contest between opposing designers in which bigger guns were fitted to penetrate thicker armour, which was then made thicker still, requiring yet larger guns, and so on.

So much has been written about the campaign in North Africa that I decided that no fresh lessons could be learned by retelling yet again the stories of Operations 'Compass' and 'Crusader' or the Battle of Gazala. Instead, I have chosen to examine the action Tel el Aqqaqir in some detail, partly because it was the climactic event of the Second Battle of Alamein, and partly because, as a breakthrough/breakout battle, it attracted considerable study while planning was in progress for the land warfare phase of the recent Gulf War.

On the Russian Front, the pendulum swung irrevocably against Germany in 1943. Hitler's attack on the heavily fortified salient at Kursk produced in the greatest tank battle in history, but the only result was to lose Germany tanks she could ill afford. Thereafter, neither the tactical expertise shown by the Germans in the bitter actions around Kharkov and in numerous cauldron battles, nor the horrific casualties sustained by the Russians, could prevent the huge Red Army grinding its way onwards into Central Europe.

While mentioning some of the difficulties experienced in conducting armoured warfare in Normandy, I also reached the reluctant conclusion that including the oft-told stories of 'Epsom' and 'Goodwood' would merely emphasise these and add little else. On the other hand, assault engineering was an area in which the British Army excelled and the story of 79th Armoured Division, otherwise known as Hobo's Funnies, from D Day until the end of the war in north-west Europe, is certainly a subject that repays study.

Italy was another area where terrain inhibited the free use of tanks. Here, as I have indicated in describing one of the actions required to break through the Gothic Line, they became first among equals in closely integrated all-arms assault teams, perhaps to a greater degree than in any other theatre of war.

Whereas in World War I the armoured car had been uncompromisingly regarded as a fighting vehicle, during World War II its most potent weapon was its radio and I have included a chapter to illustrate the point.

Once the Normandy Campaign had been concluded the scale of the US Army's contribution ensured that it became the senior partner in the land war in western Europe. Tempting as it was to describe American armoured divisions cutting their swathes across France and Germany, this would, given that cumulative experience is the theme of this book, simply have duplicated much that has been said above. Conversely, any army under pressure reveals its real worth, and this was true of the US Army during the first days of the Battle of the Bulge. There is much that can be learned from the series of checks imposed by small teams of tanks and tank destroyers on the German spearheads approaching Bastogne, the combined effect of which was to win the time in which this vital road junction could be secured against capture.

For British, Australian and American tank crews alike, fighting the Japanese presented problems that were not encountered anywhere else. It was not simply that much of the fighting took place in difficult jungle terrain. The Japanese fought to the death as a matter of course, refused to be intimidated by tanks and, under equipped as they were for mechanised warfare, developed suicidal methods of anti-tank warfare that could only be defeated by good tactics and steady nerves.

The question of individual achievements by tank commanders and crews is one upon which opinions are divided. During World War II the Germans initiated the concept of the tank ace. This is not an idea which found support in either the British or American armies. Nevertheless, it can hardly be denied that men of outstanding courage and ability served in both world wars. Some, like Lieutenant Clement Arnold in his Whippet *Musical Box*, or Count Hyazinth Strachwitz, the master of the tactical battle on the Russian Front, added much to our store of accumulated knowledge, and for that reason alone deserve a chapter to themselves.

I have devoted the last three chapters to post-World War II subjects. The Sinai Campaign of 1967 saw the Israelis, using Western equipment and ideas, pitted against the Egyptians, who employed Soviet equipment and tactical doctrine. It confirmed not only that the Blitzkrieg technique retained its validity on a modern battlefield, but also that the Soviet Kursk-style concept of defence in depth was not necessarily suited to every situation. In Vietnam the French, starved of the means to wage war by their own government, fought a hopeless battle against an enemy who was a master in the art of ambush. When the Americans and their allies entered the conflict, they brought with them modern vehicles, immense firepower and new ideas, with the result that both they and the Australians won their tactical battles. Operation 'Desert Sabre', the land battle to liberate Kuwait, once again demonstrated Western technical and tactical superiority, bringing together in its execution many identifiable strands from earlier conflicts. It also gave some hint of what might have happened if the Warsaw Pact had launched an attack on NATO. The generals, so often maligned for planning to fight the war they have just finished, were seen to have planned for that to come; and in this, of course, they had been assisted by 90 years of consolidated experience and two generations of incessant manoeuvres along the now-defunct Iron Curtain.

Bryan Perrett, March 1994

In the Beginning....

Mechanised land warfare as we understand it today began not with the invention of the tank but with the introduction of the armoured car, which has a much longer ancestry. Indeed, it is possible to argue that in remote classical times many of the armoured car's roles were performed by the chariot which, in middle eastern armies, was crewed by a driver, a soldier who fought with bow or javelin to kill his opponents from a distance, and sometimes a shield bearer who provided a degree of protection for everyone. So prestigious was the chariot as a weapon of war that the status of kings was measured by the number they could field.

As the horse evolved to the point at which it could carry an armed man, so the importance of cavalry increased while that of chariotry declined, save in a remote, backward island lying off the western coast of Europe, known to the Roman world as Britannia. When, in 55 and 54 BC, Julius Caesar carried out his two reconnaissances in force to the island, his troops found themselves confronted, inter alia, by swarms of light chariots. Such an encounter was unexpected and at first the legionaries found it very unsettling, for while the chariots were driven at reckless speed along their ranks, the warriors among the crews would perform the unbelievably daring feat of venturing along the pole between the horses to hurl their spears to best effect. Nor was this all, for if a gap appeared in the Roman ranks, chariots would instantly converge on it and the warriors would dismount to fight on foot in the hope of exploiting the advantage, while the drivers remained close by to pick them up in case of trouble. In a sense these chariots were performing the role of embryonic personnel carriers although the point is largely philosophical.

Yet, long before the dawn of the Middle Ages, the chariot had become a distant folk memory. The battlefield was dominated by armoured knights, although even they could be worsted by such formidable opponents as the English longbowmen, Swiss pikemen and the Hussite wagon forts. The last, in which wagons armed with artillery were drawn up in a circle, generally on a hilltop, are of great interest. However it would not be altogether correct to regard them as an evolutionary step in the development of mechanised warfare as they were simply the only practical means of defence the Hussites could devise against their fully armoured enemies.

It was the gunpowder revolution which finally ended the dominance of the knight. Thereafter, tactics went into the melting pot until the new means of waging war were fully understood. By the end of the seventeenth century a balance had been established between infantry, cavalry and artillery, and this was to be maintained for the next two hundred years. Nevertheless, the concept of a vehicle which provided protection for its occupants and enabled them to engage the enemy from within retained a hold on the imaginations of military thinkers throughout successive periods of change, their principal and apparently insoluble problem being the means by which it was to be powered.

Over the years various ingenious ideas were put forward, none of which was very practical. In 1335 an Italian physician named Guido da Vigevano proposed a sort of cart fitted with two windmills, the power thus generated being transferred direct to the roadwheels by an exposed wooden gear train. How the vehicle was to be steered and braked or the drive disconnected is uncertain, but the most obvious disadvantage of the arrangement was that a wind of suitable strength and direction could not be guaranteed when needed.

About 150 years later the great Leonardo da Vinci sketched his ideas for an armoured fighting vehicle, a scale model of which can be seen in the Tank Museum at Bovington in Dorset. In appearance the vehicle resembled a large bowl surmounted by a parasol, both being covered by iron plates, with spaces for guns between the two. It travelled on four wheels which emerged from its belly and is unmistakably an armoured car. Within, the crew were to turn cranks from which the power would be transferred to the wheels through rudimentary gears. With such a low power to weight ratio exhaustion would have set in very quickly. At best the vehicle's speed on hard level going would have been terribly slow, and on soft going its high ground pressure and limited clearance would have ensured that it bogged down at once. Again, it is difficult to see how the vehicle could have been steered. Da Vinci's design, however, was far more advanced than that of da Vigevano.

There the matter had to rest for nearly 400 years, until steam provided the first reliable means of self-propulsion. In 1854 a certain James Cowan, described for some reason as a 'philanthropist', had the idea of enclosing a Boydell traction engine within a domed iron cover rather like an upturned soup tureen, pierced by several ports through which small cannon protruded. Both Boydell and Cowan had a clear understanding of the ground pressure factor and the vehicle retained the original traction engine's 'feet' - short reinforced boards fixed to the circumference of the road wheels to spread the weight over a wider area - thereby contributing unconsciously to the development of the linked track. Less satisfactory was the means of steering, which relied on one small pilot wheel. Worse still, Cowan had discovered the classic predicaments faced by every subsequent fighting vehicle designer, namely ergonomics and the demands made by the machine's different functions on the strictly limited internal space available. Not much thought seems to have been given as to how furnace, boiler, flywheel, assorted machinery, coal, breech-loading guns, ammunition, driver and gunners could be accommodated, nor how the human element of the equation, roasting between the armour plating and the boiler, might perform. It was almost certainly for these reasons that the Select Committee appointed by the then Prime Minister, Lord Palmerston, rejected the design. It is also possible that Palmerston himself may have dealt the project its death blow when, observing that the public-spirited Mr Cowan had added a number of enormous scythes to the exterior of the vehicle, he described it as barbaric. Enraged, Cowan purchased space in the popular press to lambast the Select Committee, whom he described as 'washed out Old Women and Senile Old Tabbies', adding a few benevolent words on the financial rewards that could be 'obtained by wading ankle deep in blood all over the world'. At this point, perhaps fortunately, he leaves the story although, once again, something had been learned.

Steam also provided another cul-de-sac along the automotive fighting vehicle's development road. Railways enabled huge armies to be mobilised, transported into the war zone and maintained in the field, and in such roles they were of vital importance.

They were, however, of very limited value in the fighting itself for, while armed and armoured trains were to become commonplace, their routes were entirely predictable, they were easy to ambush, and could literally be halted in their tracks by the removal of a single section of rail. Consequently, in wars between regular armies their employment was purely local and restricted to the most favourable circumstances, although in colonial and guerrilla warfare, where there were often vast hinterlands to be patrolled, they had a wider application.

The boundaries of science were now being expanded at a bewildering pace. In 1885 Gottlieb Daimler and Karl Benz both produced internal combustion engines. It was inevitable that, since the power unit of the internal combustion engine was so much smaller than that of steam-driven vehicles it would be increasingly applied to road transport. In 1895 an American entrepreneur, Edward Pennington, best described as a smart operator whose diverse interests included the construction of airships and motor cars, arrived in the United Kingdom. The following year he produced a practical design for an open topped car mounting two machine-guns. This had the approximate shape of a bathtub surrounded by sloping armour skirts, with shields for the machine-guns. There was no suspension as such, but Pennington had developed a patent pneumatic tyre, then regarded as an incredibly advanced feature, and he suggested that a set of these would compensate for the hardness of the ride. Unfortunately, the design attracted little interest and was never built.

In 1899 a British inventor, Frederick Simms, produced the world's first self-propelled fighting vehicle, which he called the Military Scout. Unlike Pennington, Simms was a thoroughly respectable professional engineer who took a pride in personally demonstrating his products, dressed in well brushed bowler hat, carefully pressed suit and immaculately polished shoes. Despite this, his ideas on the subject were not altogether original. The Motor Scout, in fact, was a quadricycle - in simple terms two bicycles joined together in parallel - and it owed much to a quadricycle mounting an air-cooled Maxim machine-gun, produced by G. H. Waite, an employee of the Humber Company in 1888. Nonetheless, while Waite's machine had required three men to pedal it along, the Motor Scout was driven by a 1.5hp De Dion engine. Armament again consisted of an air-cooled Maxim, mounted on a tripod ahead of the handlebars, with an ammunition tray below. The vehicle could undoubtedly have performed the function for which it was built, albeit briefly, as protection was restricted to a small gunshield. To Simms' disappointment, the War Office showed not the slightest enthusiasm for the idea.

He tried again after the outbreak of the Second Boer War and in 1901 he borrowed Pennington's bathtub concept to produce a small, armoured, petrol-driven rail car armed with automatic weapons and a searchlight, which is believed to have been used to patrol the railways in Kenya. Thus encouraged, Simms designed and arranged for Vickers to build a much larger road version that can·justifiably lay claim to being the world's first true automotive armoured car. Named the War Car, it was armed with two water-cooled Maxims at the front and one Vickers-Maxim one-pounder Pom-Pom at the rear. A large vehicle, 28 feet long, 8 feet wide and 10 feet high, it was driven by a Simms-Daimler 16hp engine through a four-speed gearbox and could reach a maximum speed of 9mph. The driver, located amidships, employed a horizontal steering wheel to control direction and was provided with a sort of periscope for use when the vehicle came under fire. When exhibited at the Crystal Palace in April 1902 the War Car provoked an enthusiastic response from the press, the general public and a group of specially invited dignitaries. Pointedly, the War Office declined to send a representative and Simms, finally acknowledging defeat, turned his attention to more rewarding projects.

Abroad, a partially armoured car built by the Société Charron, Giradot et Voigt was tested by the French Army in 1903 but was not accepted. In 1904 the Charron company produced two fully armoured, turreted cars, one of which was accepted by the French Army and despatched to Morocco, while the second was purchased by the Tsar's government and used against rioters in St Petersburg to such good effect that further orders were forthcoming. A similar car built by Austro-Daimler was successfully demonstrated at the annual manoeuvres of the Imperial German and Austro-Hungarian armies in, respectively, 1905 and 1906, but was adopted by neither. The Germans, however, did express interest in a different sort of vehicle, examples of which were exhibited by the Ehrhardt organisation in 1906, and then by Daimler in 1909. This was known as the Ballon Abwehr Kanone (BAK) or anti-balloon gun, and was intended for use against airships, the essence of the design being simply a light lorry mounting an anti-aircraft gun and ammunition locker, protected overall by armour plate. The Italian Army was to be the first to commit armoured cars to an active service environment when in 1913 it despatched a small number of turreted Isotta-Fraschini and Fiat cars to Libya. This vast territory had been won by Italy during her war with the Ottoman Empire the previous year and since the inhabitants, notably the powerful Senussi religious brotherhood, were violently opposed to the idea, it was proving very difficult to bring it under effective control. While comparatively little is known about the operations in which the cars were involved it seems probable that, as the Italian presence did not extend far beyond the coastal strips around Tripoli, Benghazi and Bardia, they were used to provide escorts for convoys moving between these areas and isolated outposts in the hinterland. The only other country to show sustained interest in the armoured car's potential was Imperial Russia, which had not only continued to purchase cars abroad, including one from the Armstrong-Whitworth company of Great Britain, but had also begun to build her own, doubtless with one eye on internal security.

Elsewhere, the attitude of general staffs to armoured vehicles in general was uniformly chilly, and for this they have been sharply criticised by generations of historians. It is, after all, part of a general officer's function to interpret the recent past in order to project the probable course events will take in a future war, and the previous fifty years had witnessed a battlefield revolution involving the introduction of the magazine rifle, the machine-gun, quick-firing artillery, barbed wire and rapidly constructed field entrenchments. These factors were steadily leading to dominance on the battlefield of defence over attack and of this the senior officers of the time should reasonably have been aware. However, they would also have observed that in all the major wars of the time, including the American Civil War, the Franco-Prussian War, the Russo-Turkish War of 1877, the Second Boer War and the Russo-Japanese War, it was the attacking armies that had won the decisive victories, often albeit at the heavier cost. These conflicts led the generals to draw the wrong conclusion from their crystal balls, that similar victories would continue to be won by attacking armies organised entirely on conventional lines. It was to be one of mankind's greatest tragedies that this view prevailed at a critical moment in history, for in the years immediately prior to 1914 the domination of defence became paramount and the result was slaughter on a scale hitherto unimaginable.

As far as armoured cars were concerned, they were a novelty which presented the generals with something of a dilemma. They might, perhaps, have some application in colonial warfare, and in some countries the police might find them useful at times, but since they conformed to none of the tactical requirements of cavalry, infantry and artillery, which alone could win wars, there was no obvious place for them in a regular army's order of battle. Yet, in the British Army at least, amateur thought was ahead of professional, neither for the first nor the last time, and the benefits of combining fire-power with mobility were clearly recognised. Contained in the mess scrapbooks of Territorial infantry battalions and Yeomanry cavalry regiments are contemporary photographs of the annual training camps, and beside the carefully posed groups there is often an officer's private touring motor car, either mounting a machine-gun or adapted to tow one on a specially constructed carriage. One Yeomanry regiment was even presented with two armoured cars by its wealthy patron, the idea being that they should provide mobile fire support for the mounted squadrons. Sadly, the War Office continued to take the stuffy view that the activities of private gentlemen in their spare time was none of its concern.

War on Wheels

Very probably the great armoured car debate would have dragged its way through another decade or so, had not war broken out in 1914. The right wing of the German armies swept into Belgium behind a dense cavalry screen which the Belgians, lacking sufficient mounted troops of their own, attempted to check with fast Minerva and Excelsior touring cars. At first, officers had potted at the enemy lancers with sporting rifles, but it was soon appreciated that far greater damage could be done if the cars were armed with Maxim machine-guns. During one of these forays the popular Prince Baudouin de Ligne was killed, so the cars were fitted with armour plate and protected weapon mountings.

While the Belgian Army was undoubtedly the first to use armoured cars in action, the credit for the further continuous development of mechanised land warfare must be accorded to the Royal Navy. If such a statement, made during the closing years of the twentieth century, seems to beggar belief, then it has to be remembered that the naval construction race with Germany had made the Navy by far the more technically aware of the United Kingdom's two armed services, with the result that the Admiralty reacted to change at a speed unequalled in its history. In 1912 it had founded its own air arm, the Royal Naval Air Service, with a view to scouting for the fleet. On the outbreak of war the defence of the United Kingdom against Zeppelin airship attacks also became the responsibility of the RNAS.

In command of the squadron based at Eastchurch on the Isle of Sheppey was Commander Charles Samson, a piratical character who had already made history in January 1912 by making the first take-off from a ship at sea with a Short Seaplane launched from a ramp fitted to one of HMS *Africa's* turrets. Samson believed that the airship menace could best be contained by attacking the Zeppelin sheds at Cologne and Düsseldorf, but these lay beyond the range of the aircraft based at Eastchurch. With the approval of Winston Churchill, then First Lord of the Admiralty, he had the squadron, together with its motor transport and the officers' private tourers, shipped across to Dunkirk. Here, as demands on the aircraft were many and their numbers few, the cars were used to supplement them by carrying out reconnaissance patrols and picking up pilots who had been forced down. As yet, the northern flank of the contending armies still remained open, so that clashes with German cavalry and cyclist patrols were inevitable. After one encounter with a German car Samson had two of his cars fitted with boiler plate armour by a local shipyard and saw to it that the motorised patrols were adequately armed with machine-guns. A great innovator, he also had two of his lorries converted into armoured personnel carriers and later adapted a third to mount a three-pounder gun which could be used to support the smaller cars.

His squadron began to do extremely well, scattering the enemy's patrols, shooting up his outposts and generally making such a nuisance of itself that on this sector the Germans were thrown onto the defensive and forced to dig trenches across the roads to bar its further progress. Simultaneously, Samson's reports to Commodore Murray Sueter, controller of the Admiralty Air Department, indicated that it could do even better with purpose-built armoured cars. On 11 September Sueter sent a memorandum to Churchill recommending that fifty should be built. Churchill, a natural rebel himself, so thoroughly approved of Samson's activities that he increased the number to one hundred. This was further than their Lordships were prepared to go and, for the moment, a compromise was struck at sixty. The most suitable vehicles were the Rolls Royce Silver Ghost and Lanchester tourers, which were commandeered from all over the country, stripped of their coachwork and fitted with a strengthened rear axle, an armoured body and a revolving machine-gun turret which, at Samson's urging, the Admiralty had developed. The success of the Dunkirk squadron's three-pounder lorry also induced the development of the similarly armed Heavy armoured car, based on the Seabrook and later the Pierce-Arrow lorry chassis.

Such an expansion obviously required a controlling organisation and to this end the Royal Naval Armoured Car Division (RNACD) was formed in October 1914 with a depot based on the airship sheds at Wormwood Scrubs and headquarters located at No 48 Dover Street, off Piccadilly. Initially the division, under the overall direction of Commander F. L. M. Boothby, was conceived as consisting of fifteen squadrons, each containing three sections of four cars, plus one Heavy car, motor-cycle despatch riders and motorised support vehicles, the last including an incredibly advanced provision for the day, a radio truck. However, a shortage of armoured cars meant that five squadrons were equipped with motor-cycle machine-gun combinations, notably the Scott-Maxim, and during a subsequent reorganisation which saw the Heavy cars grouped together the number of squadrons was increased to twenty.

The RNACD caught the public imagination and no problem ever arose regarding the recruitment of suitable officers, many of whom travelled from as far afield as Australia, New Zealand, South Africa, Canada, India and other parts of the Empire to join. Wealthy individuals and groups willingly paid the cost of raising and equipping squadrons, among them the Duke of Westminster, reputedly the richest man in the United Kingdom, who was given command of No 2 Squadron and brought along many of his friends and even his personal jockey. Another such man was a dashing Unionist Member of Parliament named Oliver Locker Lampson who, with the assistance of the Ulster Volunteer Force, also financed, recruited and commanded his own squadron.

Filling the ranks was another matter, for in 1914 the only people capable of handling and maintaining a motor vehicle were a handful of private enthusiasts, professional chauffeurs, lorry and omnibus drivers, garage mechanics and members of the infant motor industry all of whom, taken together, added up to but a tiny percentage of the population. Keen to acquire their services, the Navy immediately found itself in competition with the Army's motorised Service Corps units. This was a contest it intended to win by fair means or foul and although it set very high standards it recognised the skills of successful applicants by rewarding them with the immediate rank of Petty Officer Mechanic, together with the appropriate rates of pay; the Army, with the same lack of imagination which had characterised its whole approach to the subject so far, could only offer a private soldier's remuneration. This in itself would probably have guaranteed the Navy's victory, but the RNACD's recruiting drive was also astutely planned to cream off the best men available and targeted the areas in which they were most likely to be found, namely London, Liverpool, Derby and Glasgow. In these cities the sight of the armoured cars touring the streets, a White Ensign streaming from a jackstaff behind their turrets, the by-now prestigious letters RNAS painted on the hull sides and the petty officer crews in their smartest shore rig, proved to be an irresistible lure for those who sought adventure. By no means all those who volunteered were accepted, for the first hurdle the candidates had to overcome was a strict evaluation of their technical competence; in Liverpool the examiners were provided by the Royal Automobile Club and in Derby by Rolls Royce Ltd. Those who passed then had to undergo the recruit's standard vetting procedure and for many this produced a real crisis. In the spirit of the times every man wanted to join something of which he could be proud, and if possible to do his bit before the war ended, which some experts predicted might be as early as Christmas. To them the very idea of being rejected was so unthinkable that middle-aged men and boys alike lied about their ages or concealed any ailment that was not immediately obvious. One recruit, almost sick with fear that mild myopia would cause him to fail the Navy's stringent medical examination, took advantage of the Staff Surgeon's temporary absence to learn the eye-testing chart by heart.

At this point the RNACD, having gone to considerable trouble to recruit a technical élite ran, into an altogether unexpected obstacle. The new petty officer mechanics had worked long and hard to acquire their skills, which in themselves guaranteed a good living and a respected place in their communities. They were, therefore, bluntly opposed to the Navy's demands that they should pass these on to others, on the grounds that by so doing they would devalue their own status and damage their livelihood. Sensibly, the Navy resolved the matter not by the rigid application of discipline, but by patient explanation that the age of the motor vehicle was just beginning and that the demand for their services would undoubtedly increase rather than decline as time went by.

Meanwhile, Samson's squadron, somewhat jealously nicknamed The Dunkirk Circus by the Army because of its growing assortment of home-made and purpose-built armoured cars, continued to break new ground. The Belgian Army had retired into Antwerp where it had been reinforced by a Royal Naval Division and a brigade of Royal Marine Light Infantry. Unfortunately, the old masonry forts of the city were no match for the Germans' modern siege artillery. It was soon obvious that the defence would be short lived and to assist in the evacuation of the garrison some 70 London buses were shipped to Dunkirk at the beginning of October. Samson's squadron escorted the convoy on the 90-mile run, providing a reconnaissance element which drove ahead to secure vulnerable crossroads, strong advance and rear guards, and more cars at intervals along the column. After two days on the road the convoy entered the city to receive an interested and friendly welcome from crowds of civilians lining the streets.

Following the evacuation of Antwerp the Belgians retired south along the coast as far as Nieuport, where the German pursuit was halted by British naval gunfire. Simultaneously, on the main battlefront the armies were vainly trying to turn each other's northern flank in the series of engagements which was to become known as The Race to the Sea. By the third week of October the Belgian defences had been linked to those of their Allies so that the trench lines now stretched from the North Sea to the Swiss frontier.

It was ironic that just as the RNACD was on the point of taking the field in force any prospect of employing the cars successfully on the Western Front should vanish altogether. Nevertheless, with the Belgian coast in enemy hands it was felt that the possibility of invasion had increased sufficiently for several squadrons to be stationed in East Anglia, and in March 1915 the Duke of Westminster's No 2 Squadron embarked for France, followed by Nos 5, 8 and 15 Squadrons; a little later the Seabrooks of several more units were sent to France and formed into Nos 16–18 Squadrons. Once the front had been locked up tight with trenches and barbed wire entanglements, there was very little that the cars could contribute, although the Heavies were able to perform some direct fixed-line shooting against the enemy's machine-gun posts, moving into position after dark and leaving before first light. The latter was easier said than done as the roads had already degenerated into shell-cratered mud wallows, ensuring that any car which became bogged was quickly wrecked by the German artillery counter-strike the following morning. It was in this manner that one of the Division's most distinguished members, Anthony Wilding, a recent winner of the Men's Singles at Wimbledon, met his death.

The war had now become global so that March 1915 also saw the despatch of several more squadrons to distant parts of the world. The conquest of Germany's African colonies, including Togo, the Cameroons, South-West Africa (now Namibia) and East Africa (now Tanzania) was already in hand. Elsewhere, Nos 3 and 4 Squadrons had also left England in March 1915, bound for the Dardanelles. Because of the cramped nature of the Gallipoli beachhead, only four of the cars were landed and these spent most of their time in dugouts at Cape Helles. The rest remained aboard their transports while the crews fought as dismounted machine-gun teams in the line. This was clearly a waste of resources and by the end of August all the cars had been shipped to Egypt.

Meanwhile, many senior officers were disappointed that, despite the trouble that had gone into raising them, the RNACD squadrons were not fulfilling their early promise. To their way of thinking, the good results produced in Africa were outweighed by the cars' impotence to influence events on the Western Front or at Gallipoli. This played straight into the hands of those who had believed from the beginning that the Royal Navy had no business in getting so deeply involved in land warfare, especially those of the Fourth Sea Lord, Commodore Cecil Foley Lambert, within whose sphere of responsibility the RNAS as a whole lay. As far as Lambert was concerned, the RNACD consisted of nothing but 'damned idlers', a term much used by those at home for anyone who, at any given moment, was not actually firing a weapon or being blown to pieces. Lambert had the less abrasive support of the First and Second Sea Lords and he was determined that the Division should be wound up. Starting in July 1915, therefore, arrangements were made for the squadrons to hand over their cars to the Army. The crews could also transfer if they so wished, but to this the response was mixed, for although officers retained the equivalent rank the Army was not prepared the maintain the special privileges granted to the other ranks, since this would have been unfair to its own drivers and mechanics. There were, however, three areas where Foley's writ had little or no effect, at least for the moment. The first was the section of cars in East Africa which, far removed from Whitehall squabbles, continued to soldier on regardless; the second was No 15 Squadron, commanded by Lieutenant Commander Oliver Locker Lampson, who was quite capable of running rings round Foley when it came to politicking; and the third was No 20 Squadron, formed by A. J. Balfour, recently appointed First Lord of the Admiralty following Churchill's departure in the aftermath of the Dardanelles debacle, to carry out a top secret construction and trials programme on vehicles vaguely referred to as 'landships'. Of the last two, more will be heard anon.

By a second curious irony Lambert's wholesale disbandments were followed by a campaign in which armoured cars were to play the decisive role. Following her entry into the war Turkey had begun supplying the Senussi in Libya with arms and ammunition, including light artillery and machine-guns, which were shipped across the Mediterranean by German submarines. During one such delivery a U-boat sank HMS *Tara*, an armed boarding vessel, in the Gulf of Sollum, then towed the survivors to Bardia in their lifeboats and handed them over the Senussi. Bardia lay within the new Italian colony of Libya, but the Italians were either unwilling or unable to do anything about securing the survivors' release and a British request directed personally to the Grand Senussi was declined on the ground that the prisoners had been entrusted to him as hostages by the Turks. Anticipating that a punitive expedition would be mounted, the Grand Senussi's Turkish advisors urged him to strike first by invading Egypt, where a current wave of anti-British feeling would guarantee widespread support and even a rising. The advice was accepted and on 17 November 1915 the Senussi army crossed the frontier.

After some fierce fighting its advance was halted near Mersa Matruh by a scratch formation known as Western Frontier Force. Among the units comprising this was an Emergency Squadron of armoured cars, formed from Nos 3 and 4 Squadrons RNACD, but as the winter rains had turned the ground into a quagmire the squadron saw little action. In January 1916 Nos 1, 2 and 3 Armoured Motor Batteries, also equipped with Rolls Royces, joined Western Frontier Force, enabling the by-now desert wise Emergency Squadron to be sent up the Nile to Upper Egypt, where Senussi bands which had occupied a number of western oases had begun to present a threat to the area. The newly arrived armoured motor batteries were actually the Duke of Westminster's No 2 Squadron, the members of which, little concerned by whose uniform they wore, preferred to serve together under the Duke and had transferred en bloc to the Army.

The Western Frontier Force, commanded by Major General W. E. Peyton, now felt strong enough to take the offensive and on 26 January it drove the Senussi out of their entrenched camp at Agagya, south-east of Sidi Barrani, after a hard fight. The climax of this was a mounted charge by the Dorset Yeomanry, made into the teeth of machine-gun fire, which resulted in 300 of the enemy being cut down; the Yeomanry's own casualties amounted to 58 of the 184 men taking part, but the loss of 85 horses reduced the regiment's strength by half. Deep, soft sand prevented the cars from getting forward and their part in the battle was confined to providing machine-gun support from ground mountings.

When Peyton resumed his advance towards the frontier on 9 March his infantry and cavalry marched along the coastal plain while the cars provided flank protection on the escarpment above. He had counted on obtaining a sufficient water supply at the Buq Buq wells but they were almost dry and the infantry began to suffer the torture of thirst. Early next morning, however, the cars discovered an alternative source and this enabled the advance to continue. Halfaya Pass and Sollum were reoccupied without the need for fighting on 14 March.

Aerial reconnaissance revealed that the Senussi had retired across the frontier and established themselves in a camp at Bir Wair. Peyton had no qualms about ordering the Duke's squadron to pursue them, as Italy was now an ally of the United Kingdom and the Italians were unlikely to resent further punishment being administered to their unruly subjects. Beyond the frontier the going was good, hard and level so that the cars were able to maintain a high average speed. Even so, their approach had been spotted by the Senussi who, abandoning their camp in such haste that the fires were left burning, took up a rocky defensive position at Bir Azeiz, some way to the west, and opened fire on the cars with their mountain artillery and machine-guns as soon as they were in range.

Once the enemy's position had been identified, the cars shook out into line, closed their hatches and went straight for it, guns blazing. Very quickly conditions inside the vehicles, already warmed to oven heat by the sun, became almost unbearable. Further heat was blasted back by the racing engines and cordite fumes thickened the air to the point that it was barely breathable. As no consideration had yet been given to fitting expense bags, drivers also suffered the further torment of hot cartridge cases showering into their open-necked shirts. What impressed itself most on participants' memories, however, was the sheer noise of the engagement – a compound of the vehicle commander's shouts, roaring engine, their own machine-gun fire and the enemy's bullets and shell splinters clattering off the armour. The enemy gunners, who were mainly Turkish regulars, were unused to engaging moving targets, the slow flight and high trajectory of their shells making it very difficult for them to range on targets that were closing in at such speed. The cars, too, were concentrating their fire on the gunners, halting briefly in turn to obtain a better shot. As the Turks died around their weapons, the whole Senussi army broke and fled across the desert. Passing through the position, the cars shot down hundreds in the ensuing pursuit; others, seeing that flight was useless, surrendered. The squadron's own casualties amounted to two men with minor injuries, probably caused by bullet splash penetrating the visors, and a few punctured tyres.

The Duke led his cars back to Sollum the following morning, bringing with him a column of prisoners, three four-inch guns, nine machine-guns, numerous small arms and 250,000 rounds of ammunition. Peyton had the unit paraded and personally congratulated it on its remarkable achievement. In fact, for the first time in history a small force of armoured vehicles, crewed by no more than 34 men, had routed an entire army unassisted. This alone would have ensured the squadron's immortality had it not immediately become involved in another incident which caught the public imagination.

Nothing had been heard of the *Tara*'s survivors for months and most people had given them up dead; nor could the prisoners shed any light on the matter. However, in a house in Sollum a letter from Captain Gwatkin-Williams, the ship's commander, was discovered. It was addressed to the commander of the British garrison but was many weeks old and had clearly been written without knowledge that Sollum was then in Senussi hands; the *Tara*'s crew, it disclosed, were being held at a place called El Hakkim Abbyat, or Bir Hacheim.

Whether they were still there Peyton had no way of knowing, but he had to assume that they were. The problem was, where was Bir Hacheim? Twenty-six years later the entire world would know its location, but in 1916 the map of the Western Desert was largely blank paper. None of the inhabitants of Sollum could help, so Peyton returned to the prisoners taken at Bir Azeiz. At length an elderly man named Ali admitted that he

22

had been to Bir Hacheim in his youth. It was, he said, five days' camel ride from Sollum and he indicated that he was willing to act as guide.

The Duke of Westminster immediately volunteered to lead a rescue attempt and a column consisting of the squadron's armoured cars, Model T Ford tenders and motor ambulances, 45 vehicles in all, left Sollum at 0100 on 17 March and began driving westward along the track leading through Bir Wair in the general direction of Tobruk. Early the following morning a large camel caravan was spotted moving along a parallel course to the south. Upon investigation it was found to be carrying war materials for the Senussi, so the drivers were taken prisoner and the camels shot. This diversion occupied several hours and it was almost noon before the column moved off again. In response to Ali's directions it continued to drive on and on to the west. It had covered approximately 100 miles since leaving Sollum when, for no discernible reason, the old man instructed the Duke to turn south. This caused immediate alarm, for in that direction, as far as everyone knew, there lay nothing but endless desert. To the Duke's mind, the guide was either leading them into some sort of trap or simply did not know his business. By 1500, after a further 20 miles had been covered, he had had enough and halted the column. A fuel check revealed that the vehicles had just sufficient petrol for the return journey, with a small margin in hand. Clearly, to go further was to risk immobilising all of Western Frontier Force's armoured cars, and most of its mechanised transport as well.

The Duke had almost decided to turn back when Ali, standing on the bonnet of a car, pointed at two distant mounds and shouted that beneath them lay the wells of Bir Hacheim. The mounds came into sharper focus as the cars raced towards them and a number of armed men could be seen running off into the desert. More figures appeared, half-naked and skeletal, waving and cheering in cracked voices as they tottered forward. They were Gwatkin-Williams and his men, five of whom had already died of starvation. The sight was too much for the armoured car crews, who tore after the fleeing Arabs and in blind fury cut them down without mercy.

After the rescued captives had been clothed, fed and given medical attention, the column set out on the return journey. On through the evening and into the night the jolting vehicles retraced their tracks northwards and then to the east. Drivers, who had been at the wheel almost continuously for the past 24 hours, found their mental and physical stamina tested to the limit. Hardest worked of all were the armoured car drivers for, notwithstanding the elegance and noble parentage of the Rolls Royce, interior comfort there was none and they sat upon a pile of mats with a simple canvas sling for a backrest, enduring agonies of cramp as hour followed hour. When, at 2300, they reached Bir Wair, now held by an Australian Camel Corps unit, the majority dropped into an exhausted sleep in their seats. The Duke himself drove on to Sollum, where arrangements were made for the *Tara*'s survivors to be transferred to Alexandria by hospital ship, but the column itself, though rested, was unable to leave Bir Wair for two days because of a sandstorm. This gave Peyton sufficient time to prepare a suitable reception so that when it finally reached Sollum it did so to the banging of an artillery salute and the cheers of the infantry.

The news of the dramatic rescue was flashed around the world. As it had taken place on their territory the Italians stirred themselves into taking a more active part in the war against the Senussi and established a fortified post at Bardia. A number of joint operations were mounted along the Libyan coast, resulting in the capture of a large quantity of stored munitions and the severing of the enemy's seaborne supply route. In November the Senussi were driven out of the western oases in Upper Egypt by a motorised column including Nos 11 and 12 Light Armoured Car Batteries, which had taken over the cars previously manned by the RNACD Emergency Squadron.

The Grand Senussi retired with the remnants of his army to the oases of Siwa, Girba and Jarabub, some 200 miles south of Mersa Matruh. Here, in a fertile but malarial valley, lying 1000 feet below sea level and protected by a difficult escarpment, he believed that he was beyond the reach of conventional armies reliant on animal transport and an adequate water supply for their survival. The British, however, were determined to bring the war to an end and they were no longer tied to conventional means. On 3 February 1917 a motorised column, spearheaded by the Duke of Westminster's armoured cars and three light car patrols, the latter equipped with Model T Fords mounting a Lewis gun, began fighting its way down the pass descending the escarpment. The Senussi resisted stubbornly for a while, supported by their one remaining mountain howitzer, but after the gun crew had been driven off they quickly lost heart. Next day the cars, brushing aside desultory opposition, broke out of the pass and entered Siwa. The Grand Senussi, his power broken, abdicated in favour of his less belligerent cousin.

Once the Senussi had become quiescent, responsibility for the security of the Western Desert was handed over to the light car patrols. Accompanied by a team of fitters and loaded with such elementary precautions as adequate fuel and water, basic recovery equipment, plenty of spare tyres and emergency air recognition panels, they covered huge areas and, in addition to their other duties, produced detailed maps that were to be used in World War II. The Duke's cars went east to fight the Turks in Palestine or to pursue a predatory career with Lieutenant Colonel T. E. Lawrence's force of Arab irregulars; it is a tribute to the quality of Rolls Royce engineering that, after such hard usage, some were still serving with the Khartoum garrison in the 1930s.

The only other armoured car unit to achieve fame comparable to the Duke of Westminster's was that commanded by Oliver Locker Lampson. Although his patriotism was beyond question, Locker Lampson was also a professional politician who enjoyed the cachet and favourable publicity attending the commander of a naval armoured car squadron. He was, therefore, gravely worried by Lambert's decision to disband the RNACD, as this would consign him to an anonymous pool of officers awaiting postings to what would inevitably be less glamorous assignments. According to his unpublished memoirs, he was dining one evening with a senior officer from the Russian Military Attaché's office in Paris. The Russians had recently accepted an offer by King Albert of Belgium to transfer his Corps of Cannon and Machine-Gun Cars to the more fluid Eastern Front, where opportunities for mobile action still existed, and the conversation naturally turned to this. Russia was, in fact, buying all the armoured cars she could lay hands on and had recently placed large orders with the Austin organisation and other companies, so the arrival of a complete unit with trained crews was very welcome indeed. Why, asked the Russian officer, did not Locker Lampson volunteer his own squadron for service with the Tsar's armies, since disbandment was the only alternative? Locker Lampson, apparently warming to the idea, asked the Russian to submit a formal request in writing.

It is quite possible, of course, that he engineered both the meeting and the conversation, for the activities of a small British unit fighting in so alien an environment would produce a constant flow of publicity well beyond its relative importance. Whatever the truth, he received an official request from the Russian government for the services of No 15 Squadron, and this he dutifully forwarded to Lambert. It is unfortunate that no record of the Fourth Sea Lord's immediate reaction has survived; very probably, it lay somewhere between the incoherent and the unprintable. Lambert's colleagues were nevertheless inclined to treat the matter seriously. The Russians had been bitterly disappointed by the Royal Navy's failure to force the Dardanelles and it would do no harm show the flag in this way; furthermore, they had undertaken to pay all the squadron's expenses themselves. Lambert not only agreed but took the view that a single squadron was not large enough for the task. Instead, an entire division of three squadrons was to go, formed from the existing Nos 15 and 17 Squadrons and such men from disbanded squadrons as chose to volunteer. The unit was to be known as the Russian Armoured Car Division RNAS and consist of three fighting squadrons numbered 1, 2 and 3, plus a headquarters and administrative squadron which, following naval tradition, would be known as Sea Base. Locker Lampson, who could never have imagined that his intrigues would pay off so handsomely, was promoted Commander and placed in overall command.

The unit sailed for Archangel on 4 December 1915 but was unable to reach its destination because the White Sea had frozen over.

The Russian Armoured Car Division finally reached Archangel at the end of May 1916, some nine months after it had embarked at Liverpool. The personnel then entrained for the long journey across Russia to Vladikavkas, on the fringe of the Caucasus mountains, being fêted at every stop along the way. The train carrying their vehicles and equipment, including one Rolls Royce and 32 Lanchester armoured cars, one hybrid Seabrook-Pierce-Arrow and two Pierce-Arrow Heavies, three towed three-pounder guns, 44 motor-cycles, and a 45-vehicle transport echelon, at length caught up with them and they were able to continue their southward journey, driving through the Caucasus along the spectacular Military Road to Tiflis, then on to the ancient fortress of Kars and Sarikamish, which they reached in the middle of August. South of Tiflis the road degenerated into a dusty track concealing a substrata of boulders. Several cracked sumps were ingeniously repaired with a mixture of soap and jam covered with sticking plaster, or molten revolver bullets. One junior officer with mechanical problems caused much mirth when his car reached the evening camp site pulled by a team of oxen, but instead of a commendation for initiative he received a dressing down for bringing the unit into disrepute.

No 1 Squadron, under Lieutenant Commander F. W. Belt, was detached for service in North Persia, where it found little to do, while Nos 2 and 3 Squadrons prepared to cross the Mush Plain and penetrate the Kurdish heartland. As Yudenich had made it perfectly clear to Locker Lampson that he could expect no logistic support whatever from the Russians, the first priority was to establish an efficient supply system. An advanced base was formed at Keupru Keui and at intervals from the Sarikamish railhead dumps of fuel and other supplies were set up, enabling a constant flow to be maintained in the direction of the fighting squadrons. While Locker Lampson, with No 3 Squadron, operated in the hostile country beyond the Mush Plain, No 2 Squadron escorted supply convoys forward from Keupri Keui. So bad was the going at times that convoys would consume up to two-thirds of their fuel load. This, of course, seriously restricted No 3 Squadron's operational capabilities, causing Locker Lampson to order the unit's home base at Newport, Monmouthshire, to fit armour plate to half a dozen of the much lighter Model T Fords and ship them out immediately.

For both squadrons, the story of their actions in Asia Minor was one of sniping, ambush and raids during which the Kurds developed a healthy respect for the cars, especially the Heavies, the three-pounder guns of which they discovered had an unpleasantly long reach. The high point was a dawn raid carried out by No 3 Squadron and a force of Cossacks on a Turkish-held village lying at the entrance to a mountainous valley.

'The firing had begun and the ceaseless chatter of the Maxims rolled over the hills', wrote Locker Lampson. 'The village had not realised its danger until one car was only 600 yards away; and then it was a though someone had lifted a stone from it, exposing black insects in an agony of commotion. From the huts, the houses, the barns, tiny figures of men shot out until the terraces and fields were crawling with battalions in flight....I started the three-pounders forthwith shelling every exit to the village. One shot completely demolished a circular house which crumpled as if constructed of cards and seemed to vomit forth a score of men, dead, wounded and fleeing. Another brought a huge piece of rock down the cliff side into a gully and blocked all egress at this point....And immediately afterwards some of the cars got in, won a terrace of the village, and did the greatest execution.'

The village was now burning fiercely and Locker Lampson drove past it into the mouth of the valley with one of the Heavies. He had been informed by prisoners that a second village, containing two battalion headquarters and an ammunition dump, lay further up the valley, and the position of this was apparently indicated by the tops of some poplars projecting above a spur. He decided to shoot it up and placed one of his best gunners, CPO Benson, behind the three-pounder.

'His first shot grazed the promontory top and raised a feathery puff of dust, the second, the third, the fourth and the fifth and the sixth all whopped over and fell in silence on the other side. Then flew the seventh, followed in a twinkling by the eighth, and back echoed a report which shook the hills and made the air tremble where I stood....We were to learn later from prisoners, who these explosions sent flying into the Russian lines, that our shells had struck a magazine and exploded it, and that a large dump had been simultaneously fired, killing over 300 Turks.'

Quite possibly, this is the first recorded instance of a target being destroyed by indirect shooting from an armoured vehicle. The Russians were delighted with the affair and awarded a total of seventeen decorations to members of the squadron.

By the middle of September the campaigning season on this sector was over. Freezing high altitude cold and the first snows made further operations impossible and, by tacit agreement, both sides withdrew into winter quarters. While No 3 Squadron commenced the journey back to Keupri Keui, Locker Lampson began to reflect that he might have been a little too clever for his own good. The Trans-Caucasian campaign held no interest whatever for those at home and, far from being besieged by eager reporters, he had disappeared from the popular press altogether.

However, no sooner had he reached Keupri Kuei than he was informed that Romania had unexpectedly joined the Allies. He decided that he would take his unit there and started lobbying at once. Bypassing Yudenich, he put a plausible case to the Grand Duke Nicholas, Viceroy of the Caucasus. The Grand Duke had no objections and referred him to General Alexeiev, the Russian Army's Chief of General Staff. Willingly, Locker Lampson undertook the long journey to St Petersburg and thence to Mogilev, where the Imperial Headquarters were located. There the shrewd Alexeiev accurately predicted that the Romanians' early success would soon be transformed into so shattering a defeat that the already overstretched Russian Army would be compelled to extend its front to the south. His view was that better employment for the British armoured cars could be found in Galicia, but Locker Lampson was adamant that he wanted to fight in Romania and Alexeiev finally agreed that the unit should be moved to Odessa, where it could be committed as the situation demanded. Locker Lampson set off for London, doubtless anticipating the flow of favourable publicity his news would generate. He was to be gravely disappointed, for their Lordships at the Admiralty, who had already put up with more than enough from him, were far from pleased by the idea of a mere Commander transferring an entire naval unit from one continent to another without so much as consulting them. The probability is that he needed all the ample powers of persuasion at his disposal, and all the political clout he could muster, just to stay in command; and while he kicked his heels awaiting a decision on his future, salt was rubbed in his wounds by the news that his unit had been heavily engaged and won itself a considerable reputation.

Gregory, his second-in-command, had been one of the first four Royal Navy pilots to qualify for their wings. Extremely capable, he had organised the transit from the Caucasus to Odessa efficiently, and from there, following the collapse of the Romanian armies, the unit moved forward into the Dobruja, a low-lying, swampy area with few roads lying to the south of the Danube delta. Here, from 30 November until 5 January, it supported a series of counter-attacks by General Sirelius' IV Siberian Corps, mounted in the hope of halting the advancing Bulgarians. One such attack is described by Locker Lampson, who based his account on Gregory's report.

'Lieutenant Commander Wells Hood had bided his chance; it was not for nothing that he had fought already in this war with armoured cars at Antwerp, in South-West Africa, in France, Belgium and Turkey. The enemy infantry had decided to advance rapidly and expected little opposition in this quarter. Wells Hood waited until they were far ahead of their trenches and moving up in the open, and then he sprang out to within sixty yards of them before they dreamed of an attack. It was in vain they fell upon their faces or tore back to their trenches; his cars caught them clean and within the radius of hundreds of yards they were stricken and mowed down. The advance as a whole slackened, wavered, then ceased, and before many minutes the Bulgarians to a man were back under cover. Still Wells Hood pursued them, harrying the trenches. He had ordered Lieutenants Ingle and Mitchell to follow at intervals of 150 yards and these cars (Lanchesters), leaving the main road, struck across country to reach a sector of trench which Wells Hood could not cover. But the going was shocking and both cars were trapped in the slimy surface many yards from their objective. In vain did Wells Hood struggle to reach them, but the intervening fields were a morass. As he turned to reverse over a ditch his power failed him so suddenly that, gazing out of the observation hole, he saw a pool of

27

petrol on the ground whither it had flowed from a couple of bullet holes in the tank. But the car was a Rolls Royce, fitted with a spare gravity tank near the steering gear, and at once he switched this on. Before the new petrol supply could reach the engine it had stopped and the gunner, Petty Officer Vaughan, pluckily leapt out in front amid a deluge of bullets and started it up; after which the car kept solidly at its excellent work unchecked and fell back only after three hours' ferocious fighting.'

The unit remained in more or less continuous action on both banks of the Danube, regularly halting Bulgarian attacks and covering Russian withdrawals. Together, battle damage and the atrocious going so seriously eroded its vehicle strength that at one period Gregory doubted whether he would be able to field a complete squadron. However, on 19 December a welcome reinforcement arrived in the shape of the unit's rear party from Newport, bringing with them the light Ford armoureds that Locker Lampson had requested in the Caucasus. These coped with the mud better the heavier cars and gradually the unit recovered its strength. In Russian eyes the British cars had performed prodigies, operating in conditions that their own armoured car commanders had pronounced impossible, and they awarded no less than 20 St George's Crosses and 26 St George's Medals in recognition.

Once the line had stabilised the unit was withdrawn to Tiraspol, where Locker Lampson rejoined it on 15 January. He bitterly resented the fact that Gregory had stolen his thunder and promptly picked a quarrel with him. Gregory was further damned with faint praise in his official reports and, when British decorations were issued for the Dobruja fighting, Locker Lampson saw to it that his Second in Command received a mere Mention in Despatches which contrasted unfavourably with the DSO awarded to some junior officers. Not unnaturally, Gregory requested a transfer, opting for sea duty. The Admiralty, well aware of what was going on, gave him command of a sloop. Having thus vented his spleen in so unpleasant a manner, Locker Lampson became his old self again. Always attracted to high places, he began calling on the Romanian Royal Family and soon developed a close attachment to the beautiful Queen Marie.

These unattractive facets of his character apart, Locker Lampson was a remarkable leader who could always extract the best from his men, although in some ways he was inclined to be priggish. For example, while he was punctilious in seeing that the rank and file received the appropriate medals for outstanding service, he was averse to recommending officers for decorations, believing that they were merely performing their duty. Likewise, when a consignment of steel helmets arrived from England he forbade their use on the grounds that, since the Russians did not wear them, the British must be seen to be taking the same risks.

By now, everyone in the unit was sufficiently familiar with life in Russia to know that serious trouble was brewing. In March 1917 the Tsar and his administration were swept away by a popular rising in which the Army played a decisive part. From this point onwards, as discipline dissolved into hostile anarchy around them, the Russian Armoured Car Division felt that they were sitting on a powder keg that could explode at any minute.

Nevertheless, the new Provisional Government announced its determination to continue the war. A summer offensive was planned in Galicia and there Locker Lampson's unit moved at the beginning of June, its losses in the Dobruja fighting having been made good by the Russians with eight Fiat armoured cars and other vehicles. On arrival it was attached to the XLI Corps, the objective of which was the town of Brzezany. The cars were to operate on two roads, driving forward to enfilade the enemy trenches while the rest of the unit, deployed in detachments along the corps front, provided direct support for the infantry's advance with Maxim, Stokes mortar and three-pounder fire.

Following a two-day bombardment the offensive began on 1 July. The cars raced through the counter-barrage and did considerable execution in the opposing trench lines. Some sped on towards Brzezany and were only prevented from entering the town by a barbed wire and sandbag barricade. A few Russian infantry regiments advanced to capture successive trench lines, but were then forced to abandon them because others refused to leave their own trenches. Officers, and in one case a regimental priest, went over the top alone and were shot down in vain attempts to encourage their men. Enraged by the sight, Petty Officer Gardner, a burly Australian serving with one of the Maxim detachments, tore into the nearest group of Russians and flung a score of them over the parapet; to his further disgust they simply crawled into the nearest shell holes.

By the end of the day it was clear that the offensive had failed, although the enemy had not returned to those trenches from which they had been cleared. For the next few days the cars went out to prevent them doing so, as the Russians refused to come forward, and on one occasion they were engaged by a large German armoured car, almost certainly a Büssing, which was driven off by the fire of the Heavies. On 8 July XLI Corps' major supply and ammunition dump was hit by artillery fire and was destroyed in a series of huge explosions; so accurate was the enemy shelling that it seems probable that the dump's precise location was betrayed by a deserter.

The Germans and Austrians were now in no doubt that most of the troops opposing them in Galicia were little better than a war-weary, undisciplined mob. When, on 16 July, they mounted a modest counter-offensive, the Russian line collapsed at once. For ten days, commencing 21 July, Locker Lampson's squadrons attempted to plug a 20-mile gap in the front, fought rearguard actions and covered the Russian retreat, all amid scenes of hysterical terror and mounting chaos. In the small town of Podgaitse, Lieutenant Commander Ruston's squadron watched in horror as a panic grew from nothing.

'Suddenly, with screams and cries of "Nemetski cavalari!" horses and carts driven by frenzied drivers galloped into the town. Anticipating trouble, we at once endeavoured to get our cars across the road to form a barrier, but we were not quite quick enough. In vain did we plead with them, expostulate with them, curse them and even threaten them. We stood in the streets and pointed revolvers at the mob and told the terror-stricken Tovarichi that it was only the work of German agents. It was perfectly useless. One and all had seen the German lancers driving the fleeing Russian infantry before them. Wonderful imaginations they were gifted with. To give a touch of realism to the scene a travelling cookhouse, going at full gallop, collided with a loaded ammunition limber and an explosion ensued. "They're shelling us!" wailed the mob, by this time hopelessly blocked in the narrow streets. In less than ten minutes the whole town was packed tight with transport coming from all directions and trying to go in one way – away from the front. Horses and carts overturned in heaps, horses with legs broken and hoofs torn off, writhing in their dying agony – a howling, frenzied, mob. To lighten their carts and so facilitate retreat, drivers emptied their loads anywhere. Sacks of sugar, flour, bales of provender, officers' kits, cases of hand grenades, field gun ammunition, spare wheels and even overcoats and rifles were all thrown into the streets. Many drivers even cut their traces and, mounting their horses, left their carts behind. At the windows were the grinning faces of the Austrian civil population. To them it was a huge joke...'

It took four hours for Ruston's cars to make their way through the town. All they encountered beyond were a few mounted scouts, who were not inclined to press the issue.

Elsewhere, the unit's cars, operating on roads and tracks and covering the intervening ground with their fire, were generally able to inflict heavy losses on the opposing infantry and cavalry as they advanced across country in open order. Once the enemy's artillery arrived, however, it was able to engage over open sights and several vehicles were knocked out, including one of the precious Heavies. More were lost when the Russians, who invariably gave way when pressed too closely, swamped them, piling aboard in their frenzy to escape until the suspension collapsed under the weight. One Lanchester kept them off by firing continuously over their heads, but an officer and several men who boarded a light Ford armoured forced its crew to drive them to safety at pistol point.

On this sector of the front the overstretched British were the only real obstacle lying in the path of the Austro-German advance. Day after day their assistance was sought by one formation commander after another, desperate to buy time in which a coherent defence line could be re-established. The fitters worked round the clock to return battle-damaged vehicles to action and even produced a replacement Heavy by mounting a three-pounder on a Fiat lorry fitted with the armour of a wrecked Ford. It speaks volumes for the unit's morale that, entirely alone in a vast ocean of troubles, it continued to fight on despite the certain knowledge that it would be deserted by its allies at the first opportunity. There were, nevertheless, moments when it could see that it was hurting the enemy badly. Sub Lieutenant Woods, spotting a wounded Austrian limping back over a rise, drove forward with the intention of capturing him, but on crossing the crest he was confronted by an entire battalion resting by the roadside. Opening fire immediately, he killed and wounded many and dispersed the rest. Sub Lieutenant Benson, the gunnery expert, took out a machine-gun post in a church belfry at 2000 yards range; and Petty Officer Rogers shot down an enemy aircraft with a Lewis gun. Prisoners stated that the cars, whose numbers they wildly overestimated, seemed to be present whichever way they advanced, and further interrogations suggested that about 2000 casualties had been inflicted, although there was no way in which this could be verified. British casualties amounted to twelve wounded, of whom only one was in a serious condition. On the other hand, two-thirds of the unit's vehicles had been destroyed and the remainder were in urgent need of repair.

Although, as a matter of policy, the scale of the Russian collapse was concealed from the British public, the achievements of the Russian Armoured Car Division was fully reported by the St Petersburg correspondents of *The Times* and *The Daily Mirror* so that Locker Lampson became a national celebrity, as he had always wanted. The following year, while writing for *Lloyd's Magazine*, he was to gild the lily somewhat that by claiming that the enemy had ordered the armoured car crews to be brought in 'dead or alive', and that the German Kaiser had personally placed a price of 20,000 Marks on his head. From General Lavr Kornilov, now Commander in Chief of the Russian Army, he received a personal note of thanks, the Order of St Anne, and 24 St George's Crosses to be awarded at his discretion.

Once the front had stabilised again the British unit moved to its rear base at Kursk. Locker Lampson, fully aware that the political situation in Russia had become even more volatile, began sending his men on home leave in batches until all that remained at Kursk was a small care and maintenance party under Lieutenant Commander R. J. Soames.

On Christmas morning members of the local Soviet forced their way into the Kursk base and, holding the British officers at gunpoint, told Soames to hand over the cars. The other ranks, in a very ugly mood, now began emerging from their billets with the Maxims and their personal weapons and it was with difficulty that Soames prevented them attacking the Bolsheviks. Bluffing brilliantly, he told the now-frightened Commissar that the cars were useless and only at the base for repair; that he was welcome to them after Christmas, which had a special significance for the British; and that in the meantime he and his followers had better clear off if they wished to avoid a bloodbath. The Bolsheviks beat a hasty retreat, leaving a couple of sentries on the gate for the sake of face. These were regularly fed in the unit's cookhouse and, astonished by their decent treatment, they not only began saluting Soames and his officers, but also made serious enquiries about joining the RNAS! In the meantime, the crews had set to rendering the cars and their weapons useless by removing or damaging small but vital working parts.

On 8 January the Soames received more visitors demanding cars and guns, this time from the Moscow Soviet. He told them to discuss the matter with their Kursk associates, with the result that the two factions fought a street battle over the spoils, the Kursk fraternity being supported by the one car they had been able to piece together from the rest. Soames had kept his superiors informed and it was decided that rear party should be evacuated immediately. It left Kursk by train on 12 January but so unsettled was the state of the country and such was the condition of the railway system that it took two weeks to reach Murmansk, where it embarked for the United Kingdom on 1 February.

By now most of the Army's armoured car units had been absorbed into the Machine Gun Corps (Motors). Known generally but by no means invariably as Light Armoured Motor Batteries (LAMBs), they served in East Africa, Egypt, Palestine, Mesopotamia, Persia, Arabia and India. Locker Lampson's unit was transferred almost en bloc to the Army which, because of its distinguished record, took the unusual step of granting its personnel their equivalent ranks. Now known as Dunsterforce Armoured Car Brigade, shortened to Duncars, it consisted of three squadrons equipped with twin-turreted Austins and two motor machine-gun squadrons with Ford and Peerless transport. In March 1918 it arrived in Mesopotamia and became part of Dunsterforce, which, under the command of Major General Lionel Dunsterville, had been given the impossible task of preventing the Turks capturing the Baku oilfield in the aftermath of the Russian collapse in Trans-Caucasia. Locker Lampson himself, now employed in interpreting events within Russia for the British government, remained in England, but in a surprising gesture for so ruthless a man he not only wrote personally to every one of his veterans, thanking them for their past services, but also presented each of them with a fifteen shilling saving certificate, entirely at his own expense.

Before Duncars had embarked for the Middle East, sixteen of its Austins had been transferred to the 17th Battalion Tank Corps for use in France, where Ludendorff's spring offensives had almost succeeded in breaking the Allied line. By late summer the initiative had returned to the Allies and during the great Battle of Amiens, which commenced on 8 August, the 17th Battalion was allocated to the Australian Corps with the task of exploiting the anticipated breakthrough. Just how they were to negotiate the shell-torn landscape of no-man's-land and cross the maze of enemy trenches beyond was left to Lieutenant Colonel E. J. Carter, the battalion's commanding officer. The problem seemed insoluble until, on the eve of the battle, Carter discovered that the wheelbase of the cars fitted the tracks left by a tank. A hasty experiment confirmed that a tank could tow two cars across country without undue difficulty. It was also believed that the cars could be hauled over the fascines dropped into the German trenches by the tanks' assault wave to assist their own crossing.

The method worked, although it gave the crews a very bumpy, slow-motion ride over which they had no control. Beyond the trench lines the cars were released but found that the Germans, as thorough as ever, had felled trees across the main Amiens-St Quentin road, along which they were to proceed. So successful had been the assault that the Australian infantry were now some miles ahead. Carter's men, therefore, had to wait in frustrated impotence while tanks towed the large trunks off the road and pioneers chopped a way through the tangle. At length the cars caught up with the infantry and passed through their supporting barrage into the open but still enemy-held country beyond.

Carter did not return to make his report to Lieutenant General Sir John Monash, commanding the Australian Corps, until midnight. He was tired, covered from head to foot in grime and grease, but happy with the results of his day's work. Fortunately, most of his story was taken down verbatim.

'We detached three sections to run down to Framerville. When they got there they found all the Boche horse transport and many lorries drawn up in the main road ready to move off. Head of column tried to bolt in one direction and other vehicles in another. Our men killed the lot (using 3000 rounds) and left them there; four staff officers on horseback shot also.'

Most important of all, the two leading cars, commanded by Lieutenant E. J. Rollings, captured an enemy corps headquarters in Framerville, abandoned in such haste that among the papers were the complete plans of a 25-mile stretch of the Hindenburg Line. Needless to say, these were put to excellent use as the Allies continued their offensive.

Carter's report continues: 'The cars then ran down the east side of Harbonnières, on the south-east road to Vauvillers, and met a number of steam wagons; fired into their boilers causing an impossible block. Had a lot of good shooting round Vauvillers, then came back to main road. Two sections of cars went on to Foucaucourt and came in contact with a Boche gun in a wood north-east of Foucaucourt. This gun blew the wheels off one car and also hit three others. However, three of the cars got away. Two others went to Proyart and found a lot of troops billeted there having lunch in the houses. Our cars shot through the windows into the houses, killing quite a lot of the enemy. Another section went towards Chuignolles and found it full of German soldiers. Our cars shot them. Found rest billets and old trenches also with troops in them. Engaged them. Had quite a battle here. Extent of damage not known, but considerable....I went a quarter of a mile beyond La Flaque. There was a big dump there, and Huns continually coming out and surrendering, and we brought back quite a lot of them as prisoners. A party of Hun prisoners was detailed to tow back my disabled car.'

The cars, of course, were performing on a mechanised battlefield the role once carried out by light cavalry, namely the pursuit of a beaten enemy. In other theatres of war, too, it had long been accepted that they provided an ideal tool for another former light cavalry task, that of reconnaissance.

It is appropriate, therefore, to conclude this brief survey of armoured car operations in World War I with an account of an action which demonstrates the flexibility the new arm had attained. By the spring of 1918 the Turkish army in Mesopotamia had been pushed back into the north of the country. In March of that year it was decided to eliminate the Turkish 50th Division holding Khan al Baghdadi on the Upper Euphrates. This was to be accomplished with a frontal attack by Major General Brooking's 15th Division while the 11th Cavalry Brigade and three LAMBs worked their way through the hills to the west and established a blocking position on the Turkish line of retreat. The operation went entirely according to plan and resulted in the surrender of the entire enemy force. Following this the cavalry and armoured cars carried out an exploitation along the Aleppo road.

On the day prior to the British attack an aircraft crewed by Colonel E. J. Tennant, commander of the Royal Flying Corps in Mesopotamia, and Major P. C. S. Hobart of the Bengal Sappers and Miners, had been so seriously damaged by anti-aircraft fire while flying over Khan al Baghdadi that it was forced to land. The Turkish commander, Nazim Bey, sent the captives to the rear.

On the morning of 28 March the 8th LAMB, commanded by Captain Tod, passed through 'Anah and was told by prisoners that the two British officers and their escort had left just two hours previously. Tod informed higher authority, who gave him permission to pursue for 100 miles and promised that, if necessary, aircraft would be used to resupply his cars with fuel. Some thirty miles further on, the cars bore down on the party just as the escort, deciding that Tennant and Hobart were an encumbrance, were openly discussing their murder. Like every armoured car commander, Tod enjoyed being off the leash and, having rescued the captives, he took advantage of the spare mileage left on his ticket. Now deep inside enemy territory, he pressed on for a further 40 miles, correctly anticipating that he would encounter little opposition so far from the front. Among those he snapped up along the road was the head of the German Euphrates Mission and a party of high ranking Turkish officials.

The incident, demonstrating as it did the potential of armour, undoubtedly influenced Hobart's post-war decision to transfer to the Royal Tank Corps. Twenty years later it was he who so thoroughly trained the 7th Armoured Division for desert warfare, subsequently raising and training both the 11th Armoured and 79th Armoured Divisions.

The Land Ironclads

N
o one individual can claim the credit for inventing the tank, although H. G. Wells foresaw both the probable form it would take and the use to which it would be put in his short story The Land Ironclads, published in the December 1903 issue of The Strand Magazine. The vehicle he described is quite recognisable even if with a length of 100ft it was overlarge, and it was evidently powered by steam. What made it particularly interesting, however, was the fact that it travelled on a series of footed wheels, some mention of which has already been made.

The footed wheel was an intermediate step between the wheel proper and the continuous linked track. It had not taken engineers very long to appreciate that if the feet were linked together they could be made into a continuous belt or track along which a vehicle's roadwheels could travel, with power being supplied direct to the track itself by means of a toothed sprocket. The obvious progression from this was to extend the track to the whole length of the vehicle, with the result that, because the ground pressure was spread over a wide area, it possessed a far superior cross-country performance to a wheeled vehicle of equivalent size and weight. By 1906 this principle was already being applied by the manufacturers of agricultural machinery to heavy duty tractors.

In 1911 an Austrian officer, Gunther Burstyn, produced a design for an armed, tracked vehicle capable of crossing trenches. The following year Mr Lancelot de Mole, an engineer from North Adelaide, Australia, submitted a plan for a tracked vehicle to the War Office. The reaction of the respective General Staffs, Austro-Hungarian and British, mirrored exactly their contemporary attitude to armoured cars - such ideas were ingenious, but they were incompatible with any of the traditional arms, and the plans were quietly filed away to gather dust.

With the arrival of trench deadlock in the autumn of 1914, both sides committed larger and larger masses of men to their offensives, which were preceded by heavier and heavier artillery preparation, but the only results were horrific casualties for the attackers. Having also employed gas and the tactics of attrition, the German Army found that these were two-edged weapons and adopted a policy of strategic defence.

Sometimes, the view of an enemy permits us to see moments in our own history in a more rounded perspective. In his book Achtung – Panzer!, published in 1937, Major General Heinz Guderian reminds his readers that it was Field Marshal Lord Kitchener who had used modern firepower with devastating results to win the Battle of Omdurman in 1898, yet that now, as Secretary of State for War, he was permitting his army commanders to 'fight like the Dervishes', sending men with cold steel to storm entrenched, heavily wired positions held by massed riflemen, machine-guns and quick-firing artillery. Of course, every one of the contending armies was doing the same, and Guderian is simply expressing surprise that the British, with their ample experience of the effects of firepower, had fallen into the trap. Nevertheless, as he was himself to emphasise later in life, only movement could bring victory and the generals of the time, fully aware of this, strove to restore it with the limited means at their disposal.

The cause of the problem, and a possible cure, had been identified by Lieutenant Colonel Ernest Swinton, a Royal Engineer officer, as early as September 1914. Swinton was Assistant Secretary of the Committee of Imperial Defence and, as the Army would not at this period permit representatives of the Press into the forward areas, he was also carrying out the duties of official war correspondent. Having observed that the Royal Artillery was using American Holt caterpillar tractors to haul its heavy guns across broken ground, he submitted a memorandum to his immediate superior, Colonel Maurice Hankey, recommending the construction of an armoured vehicle capable of crossing trenches, breaking through barbed wire entanglements and destroying machine-gun posts. At first Kitchener was dismissive of the idea but under pressure from the government he at length consented to trials. These took place in adverse weather conditions on 17 February 1915 and were carried out by a Holt tractor towing a trailer intended to represent the additional weight of armour, guns and crew. Not surprisingly, the results were disappointing and the project was shelved.

There the matter might have rested had it not been for the Navy. As a member of the Committee of Imperial Defence, Churchill naturally received Hankey's minutes and the concept of the Machine-Gun Destroyer dovetailed with some ideas of his own and Sueter's. On 20 February 1915 he set up a Landships Committee under Mr Eustace Tennyson d'Eyncourt, the Director of Naval Construction, the members of which included Mr William Tritton, Managing Director of William Foster & Co Ltd, agricultural machinery manufacturers of Lincoln, Colonel R. E. B. Crompton, whose experience of land traction went back as far as the Crimean War, Flight Commander T. G. Hetherington and Lieutenants W. G. Wilson and A. Stern, all of the RNAS. The purpose of the Committee was to design and build the sort of vehicle envisaged by Swinton, with the practical work of trials and testing being carried out by the specially formed No 20 Squadron RNACD at Burton-on-Trent. Two further squadrons, numbered 21 and 22, were to have provided the first operational crews, but this was altogether too much for the irascible Fourth Sea Lord, Commodore Lambert, who, it will be recalled, was trying to do away with the RNACD altogether.

'Caterpillar landships are idiotic and useless', he raged. 'Nobody has asked for them and nobody wants them. Those officers and men are wasting their time and are not pulling their proper weight in the war. If I had my way I would disband the whole lot of them. Anyhow, I am going to do my best to see that it is done and stop all this armoured car and caterpillar landship nonsense!'.

Fortunately, d'Eyncourt was able to convince the new First Lord of the Admiralty, A. J. Balfour, that without No 20 Squadron the whole project would grind to an immediate halt and, thanks to pressure exerted by him it was reprieved, expanding from just 50 men in July 1915 to over 600 by the end of the war.

The work of the Landships Committee would have been materially assisted, and undoubtedly accelerated, if it had received the benefit of de Mole's ideas. Very probably, changes in personnel within the War Office's expanding bureaucracy had led to these being forgotten beyond recall since they were submitted in 1912. That, however, does not excuse the fact that when de Mole, now serving as a corporal in an Australian infantry battalion, submitted them afresh to the War Office in 1915, they were again confined to the filing cabinet. Had he sent them to the Admiralty they might have been received very differently, but such a course would hardly have entered the logical mind of a professional engineer; to many, the ways of the armed services may sometimes seem, as Churchill once said of the Soviet Union, a riddle wrapped in an enigma. In the long term, when the first generation of tanks was built, they bore so startling a resemblance to de Mole's vehicle that the question of rights arose. In 1919 de Mole was honoured by the Royal Commission on Awards to Inventors, which declared that: 'He is entitled to the greatest credit for having made and reduced to practical shape as far back as the year 1912 a very brilliant invention which anticipated and in some respects surpassed that actually put into use in the year 1916. It was this claimant's misfortune and not his fault that his invention was in advance of his time, and failed to be appreciated and was put aside because the occasion for its use had not then arisen.' A one-eighth scale working model of the vehicle, submitted by de Mole with his plans, was later presented by him to the Australian War Memorial in Canberra.

The Landships Committee perforce had to start from scratch and proceeded along two parallel lines of research which were ultimately to prove complementary. The ability to cross obstacles was a prime necessity and some believed that this could best be achieved by mounting the vehicle on large wheels. The first suggestion, involving three 40-foot diameter wheels arranged in tricycle form, was quickly scaled down, but even with fifteen-foot diameter wheels it was considered to be too large. The second line of research relied on application of the continuous track system. Two Bullock Creeping Grip tractors were imported from the United States and joined by an articulated coupling to provide the desired trench crossing capability. As this arrangement proved clumsy and the coupling was too weak, it was decided to build a new vehicle half the total length of the articulated version, and fit it with lengthened Bullock tracks. When completed in September it was given a box superstructure and a hinged two-wheel tail that, when lowered, acted as an aid to steering in the manner of a boat's rudder. The only major problem encountered with the vehicle, known as the No 1 Lincoln Machine, was that the Bullock commercial tracks were too narrow for the additional work they were required to do, and also had a tendency to shed. Undeterred, Wilson and Tritton set too and redesigned the track and suspension units. The rebuilt vehicle, named Little Willie, was completed in December and had a much superior performance.

While these developments were taking place, Lieutenant Colonel Swinton, whose memorandum had set the whole chain of events in motion, had succeeded in reviving the Army's interest in the project. The difficulty was that Little Willie had been designed to surmount a parapet two and a half feet high and cross a trench five feet wide, while the Army wanted a vehicle that could surmount a parapet four and a half feet high and cross a trench eight feet wide. Wilson's answer to the apparently insoluble conundrum was brilliant. He decided to retain the hull of the No 1 Lincoln Machine, and to reintroduce the big wheel principle by carrying the tracks right round the vehicle. The real stroke of genius, however, lay in compressing the shape of the wheel into a rhombus so that the bottom run was shaped like an arc taken from a 60-foot diameter wheel, while the vertical step from the ground to the forward track horns was more than adequate for surmounting parapets and crossing trenches. Armament consisted of two naval six-pounder guns mounted in sponsons on either side of the hull to reduce height, and four Hotchkiss machine-guns. Like Little Willie the new vehicle, known variously as the

Wilson Machine, HMLS (His Majesty's Landship) Centipede and Big Willie but more commonly as Mother, was powered by a Daimler 105hp sleeve valve engine and fitted with trail wheels.

Work on Mother had begun even before Little Willie was completed. In January 1916 she was herself ready for trials and was taken to the Marquess of Salisbury's estate at Hatfield Park in Hertfordshire, where an obstacle course had been laid out, simulating crater fields, ditches, streams, wire entanglements and German trenches, one of which was actually wider than any thus far encountered. On the 26 January, driven by Chief Petty Officer Hill of No 20 Squadron, she coped with these to the satisfaction of the Landships Committee, and did so again on 2 February before a distinguished audience which included Kitchener, Lloyd George, then Minister of Munitions, members of the Cabinet, the Army Council, the Admiralty and General Headquarters in France. A week later Hill gave a further demonstration for King George V, who also travelled in the vehicle for a while.

Kitchener's reaction was gruff and lacking in encouragement, but it was the opinions of the GHQ officers that counted for most and they were enthusiastic. On 12 February the first order for 100 of the vehicles was given to Foster's and the Metropolitan Carriage and Wagon Company of Wednesbury. Half were to be 'males', perversely carrying the same armament as Mother, and half were to be 'females' armed with six machine-guns, four of which were mounted in smaller side sponsons, the idea being that these would prevent the males being swamped by enemy infantry; later, a 'hermaphrodite' version was to be produced, with a male sponson on the right and a female sponson on the left.

It might seem strange that a weapon system that was to change the nature of twentieth century land warfare should have been given the uninspiring and apparently meaningless name of 'tank', yet that was its precise intent. Because of the secrecy with which the whole project was shrouded, the purpose behind the unusual activity at Foster's had to be given some form of cover. The term landship provided too obvious a clue, but there was nothing unusual about a company manufacturing agricultural machinery being asked to produce a mobile water cistern or tank to a military specification; there were, too, some very broad hints that, when completed, the vehicles would be shipped to Russia. The name stuck and has been in common usage ever since throughout the English speaking world.

It had taken just twelve months to create an entirely new sphere of military technology, a breathtaking achievement when one considers the years of work required to produce a modern tank design with all its complexities. The sheer speed involved in getting the Tank Mark I off the drawing board and into production virtually guaranteed that it would incorporate a number of undesirable features. The most obvious of these was that no less than four men were required to steer the vehicle as a team effort – the driver to declutch and lock the differential, the commander to brake the neutral track and two gearsmen to operate the secondary gears on each track; in this context the trail wheels were simply another complication providing minimal assistance and, as they were vulnerable to shellfire, they were soon dispensed with. A particularly dangerous feature of the design was the petrol tank, located inside the vehicle, adding greatly to the risk of fire; furthermore, the engine depended upon a gravity feed system so that when the vehicle was in a steep nose or tail down position it could be starved of fuel at critical moments.

A total of 150 Mark Is were constructed. Marks II and III, of which 50 of each were built, entered service respectively in January and February 1917; they incorporated a number of minor modifications and the Mark III was fitted with improved quality armour that was proof against the German anti-tank rifle. The Mark IV, which began reaching the front in April 1917, provided a number of improvements including an external fuel tank mounted at the rear, a vacuum-feed petrol system, an exhaust silencer, a cooling and ventilation system, an unditching beam stowed on rails running the length of the roof, and maximum armour thickness increased from 12mm to 14mm, all of which stemmed from user experience in the field; altogether, the number of Mark IVs completed amounted to 420 male and 595 female tanks. The Mark V arrived in France in May 1918. It represented a major step forward in that it incorporated an epicyclic steering system designed by the now Major Wilson, enabling the vehicle's movements to be controlled by the driver alone. It was also driven by a more powerful 150hp Ricardo engine that increased the maximum speed available from 3.7mph to 4.6mph. In addition to the 400 standard Mark Vs completed in equal male and female proportions, a further 200 males and 432 females, designated Mark V*, were built with an extended hull that increased trench crossing capacity.

These vehicles provided the British Army with its armoured striking force, but long before they could be sent into action it was necessary to set up a complete new arm, with every consideration that this entailed. Swinton was appointed its overall commander, and the field commander in France was to be another Royal Engineer officer, Colonel Hugh Elles. The latter's small group of staff officers, selected for their enthusiasm and intelligence, included two men who were able to interpret the rapid pace of technical advance and foresee that, in the not-too-distant future, the tank would become much more than a tool of trench warfare. They were Captain G. Le Q. Martel, the Brigade Major, who predicted that battles between opposing tank forces would become inevitable, and Major J. F. C. Fuller, the Chief of Staff, who went even further by proposing that tanks should be used specifically against the enemy's command centres, thereby eliminating his ability to fight.

The first problem to be confronted by the new arm was its title. At first it was simply known as the Tank Detachment, but for security reasons this was quickly changed to the Armoured Car Section of the Motor Machine Gun Service. In May 1916 it became the Heavy Section Machine Gun Corps and in November the Heavy Branch Machine Gun Corps. Not until June 1917 did it become the Tank Corps. To look some way into the future, in 1923 George V honoured the Corps with the prefix Royal, and in 1939, on the formation of the Royal Armoured Corps, the Royal Tank Corps became the Royal Tank Regiment, which title it has held ever since.

A further question demanding an immediate answer was who was to train the instructors who were to instruct the crews? No 20 Squadron RNACD could provide limited assistance, but it was fully occupied with other matters. Some of the primary driving instructors were drawn from the Service Corps' Caterpillar Companies and the Heavy Siege Batteries of the Royal Garrison Artillery, who were already familiar with the workings of tracked vehicles. Others were recruited by the editor of the *Motor Cycle* magazine, Mr Geoffrey Smith, from among his contacts in the motor manufacturing industry. More still were recruited by combing the Army for volunteers.

Training the gunnery instructors was a much simpler matter. As the six-pounder was a naval weapon the Royal Navy provided a course for them at its own Gunnery School on Whale Island off Portsmouth. This included a day or two at sea during which the weapon was fired from the pitching deck of a destroyer to simulate the sort of conditions that would be encountered in action.

Raising the first crews also presented its problems. The Corps' first home was at Siberia Camp, Bisley. Nearby was the depot and training school of the Motor Machine Gun Service, and from this some 700 officers and men were transferred in March 1916. As for the rest, Swinton was again forced to rely on volunteers. It might be thought that the recently disbanded RNACD squadrons would provide an ideal source, but when he visited them the need for secrecy prohibited him from telling them anything more than that he was seeking men 'for an exceedingly dangerous and hazardous duty'; and, of course, that on transferring the POs would serve as privates with suitably reduced rates of pay. Naturally enough, though a number of officers came forward, the rest were not inclined to take so unpromising a leap in the dark. Nevertheless, volunteers did appear from all over the Army, some drawn by the prospect of danger, some by veiled references to the technical nature of the arm, and some because they believed that they could contribute more to the war effort in this way. At first their only common item of uniform was the tank arm badge, still worn by the Royal Tank Regiment, which was initially awarded as a brevet, rather like a pilot's wings, for those who had qualified in their crew trades. Long after other ranks had been issued with the crossed Maxims of the Machine Gun Corps as a cap badge, officers continued to wear the uniform of their parent regiments. This diversity was reflected in the origins of the officers appointed to command the first six tank companies to be formed – two came from the RNACD, one from the Royal Artillery, one from the South Wales Borderers, one from the Cameronians and one from the Argyll and Sutherland Highlanders.

These six companies, lettered A to F, each consisted of 25 tanks. In January 1917, as the Corps grew in size, they were expanded into battalions each with three companies of 25 tanks each, and further battalions were formed. To avoid confusion in this and later chapters, their lineage followed the pattern – A Company – A Battalion —1st Battalion Tank Corps/RTC – 1st Royal Tank Regiment (1 RTR).

Before describing the part they were to play in the war, it is necessary to say something of the appalling conditions in which the tank crews fought. Such intense heat was generated by the engine that the men wore as little as possible. The noise level, a compound of roaring engine, unsilenced exhaust on the early Marks, the thunder of tracks crossing the hull, weapons firing and the enemy's return fire striking the armour, made speech impossible and permanently damaged the hearing of some. The hard ride provided by the unsprung suspension faithfully mirrored every pitch and roll of the ground so that the gunners, unaware of what lay ahead, would suddenly find themselves thrown off their feet and, reaching out for support, sustain painful burns as they grabbed at machinery that verged on the red hot. Worst of all was the foul atmosphere, polluted by the fumes of leaking exhausts, hot oil, petrol and expended cordite. Brains starved of oxygen refused to function or produced symptoms of madness. One officer is known to have fired into a malfunctioning engine with his revolver, and some crews were reduced to the level of zombies, repeatedly mumbling the orders they had been given but physi-

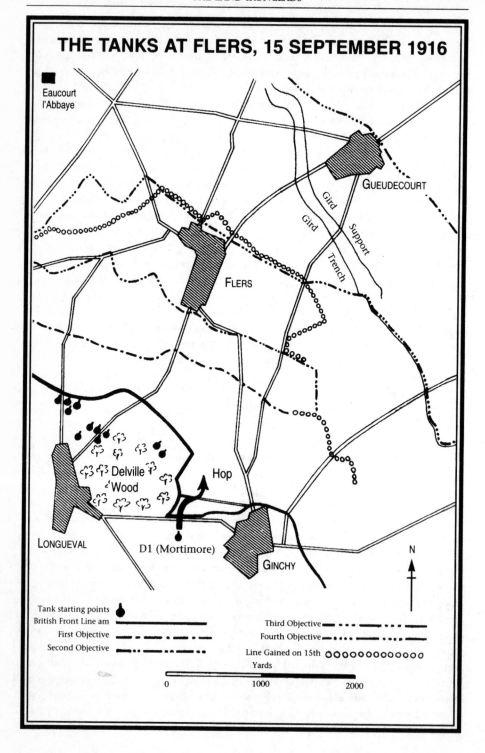

THE TANKS AT FLERS, 15 SEPTEMBER 1916

Eaucourt
l'Abbaye

GUEUDECOURT

Gird

Gird

Support

Trench

FLERS

Delville
Wood

Hop

LONGUEVAL

D1 (Mortimore)

GINCHY

N

Tank starting points
British Front Line am
First Objective
Second Objective

Third Objective
Fourth Objective
Line Gained on 15th

Yards

0   1000   2000

cally unable to carry them out. Small wonder then, that after even a short spell in action, the men would collapse on the ground beside their vehicles gulping in air, incapable of movement for long periods.

In addition, of course, there were the effects of the enemy's fire. Wherever this struck, small glowing flakes of metal would be flung off the inside of the armour, while bullet splash penetrated visors and joints in the plating; both could blind, although the majority of such wounds were minor though painful. Glass vision blocks quickly became starred and were replaced by open slits, thereby increasing the risk, especially to the commander and driver. In an attempt to minimise this, leather crash helmets, slotted metal goggles and chain mail visors were issued, but these were quickly discarded in the suffocating heat of the vehicle's interior. The tanks of the day were not proof against field artillery so that any penetration was likely to result in a fierce petrol or ammunition fire followed by an explosion that would tear the vehicle apart. In such a situation the chances of being able to evacuate a casualty through the awkward hatches were horribly remote.

Despite these sobering facts, the crews willingly accepted both the conditions and the risks in the belief that they had a war-winning weapon. The first units to leave for France, C and D Companies, were anxious to get into action to prove their point. Ideally, to obtain the maximum effects of surprise and shock, the existence of the tanks should have been concealed from the enemy until many more were present, and then they should have been committed en masse. However, while Field Marshal Sir Douglas Haig, Commander of the British Army in France, was aware of this, the Battle of the Somme had been raging since 1 July without any tangible return for the huge casualties that had been incurred. Now, two months later, he felt that he must use every means in his power to achieve the desired breakthrough, so both companies were detailed to lead an attack on the Flers-Courcelette sector of the front. As far as concentration was concerned, Haig had to contend with the competing demands of his subordinates and allocated seventeen tanks each to XIV and XV Corps on the flanks of the attack, eight to III Corps in the centre, and seven to the Canadian Corps. This, in effect, meant that the tanks would be committed in penny packets of two or three right along the line.

The operation commenced at 0515 on 15 September with a preliminary attack by three tanks and two companies of the 6th King's Own Yorkshire Light Infantry against an advanced German trench known as Hop Alley, to the east of Delville Wood. One tank broke down immediately and a second became ditched, so the third, *D1* commanded by Captain H. W. Mortimore, a former Royal Marine Artillery officer who had served with the Naval Division at Antwerp and later with a mobile anti-aircraft battery in the Ypres Salient. Aged 23 at the time, he was the only member of the crew who had been in action before, and his men were even younger. Over 50 years later, a quiet man now living in retirement, he recalled his experiences for the late Major John Foley, who included them as a postscript to his excellent book *The Boilerplate War*.

Mortimore remembered that while *D1* was moving ahead into no man's land the KOYLI followed him, using an old communication trench for cover. There were flashes along the edge of Delville Wood, indicating the presence of enemy machine-guns, and he ordered his six-pounder gunners to engage them. As friendly shellfire was also falling on the objective it was difficult for him to observe the results, but he believed that his fire had had some effect for the KOYLI charged past with fixed bayonets and stormed their way into the wood. He positioned *D1* astride a nearby trench and began to rake it with his machine-guns, noting the expressions of shock and horror on the enemy's faces as they tumbled up from their dugouts.

The main attack was timed to commence at 0620 and Mortimore's orders were to join in as it reached him. He did so but had only covered a further 300 yards to the east when a direct hit from a shell smashed in the starboard sponson, killing two of the crew, and broke a track. Though now immobile, *D1* continued to engage targets in the enemy trenches until the infantry again overtook her, but after that Mortimore and his men could only sit and wait until such time as the vehicle could be recovered. The site of their action, the first ever fought by a tank crew, is today marked by the Tank Corps Memorial.

As a result of mechanical problems, only 32 tanks were available to lead the main attack, and of these a further nine broke down immediately and five became ditched. The fortunes of the remainder were varied. A number fell victim to artillery fire controlled by observers in tethered balloons, others wallowed to a standstill in boggy going, some reached the enemy lines but had to retire because their infantry were pinned down, and a scattered few took their objectives in conjunction with the infantry. Most drivers had never seen a battlefield before and found the greatest difficulty in overcoming a natural reluctance to drive across the numerous British and German bodies strewn across their path.

The best results of the day were obtained at Flers village, against which something like a concentration of armour had been achieved. Here *D6* (Lieutenant Legge), *D16 Dracula* (Lieutenant Arthur Arnold) and *D17 Dinnaken* (Lieutenant Stuart Hastie) closed up to the village and suppressed the opposition so effectively that the infantry were able to take possession of the ruins, none of which stood more than shoulder height, at about 0800. Arnold immediately sent off a carrier pigeon to inform the infantry corps headquarters that the objective had been taken, this being the only means by which tanks could communicate information quickly to senior officers in the rear areas. He and his crew then set about cooking breakfast and replenished the fuel tank from tins carried on the trail wheels. After this they became involved in an unsuccessful attempt to extinguish a fire in another tank nearby. From time to time Arnold moved his own vehicle, which had evidently attracted the attention of one of the enemy's observation balloons. During the afternoon he assisted the infantry in beating off a counter-attack but withdrew when *Dracula* became the target of a field gun firing over open sights. The infantry in Flers had now been reinforced and in the fading light he returned to base, harried by artillery fire for much of the way.

An observer in a British aircraft had watched Hastie's *D17 Dinnaken* (Don't know in Scottish vernacular) break into the village and he sent back the following report: 'Tank seen in main street of Flers going on with large number of troops following it.' This was immediately transposed by the popular press into truly sensational news: 'A tank is walking up the High Street of Flers with the British Army cheering behind!'. The truth, painstakingly researched by John Terraine, was more prosaic. It had been necessary for Hastie to flog his engine without mercy to get to Flers at all and by the time he arrived the big ends were knocking hard. After passing through the village he returned to the main street and halted to ask an East Surrey officer directions to the hamlet of Gueudecourt, which was to be the final objective. At that moment the East Surreys were winkling prisoners out of dugouts and cellars and forming them into a column of fours ready to be marched to the rear, so the probability is that this was the 'large number of troops' noted by the air observer. Hastie continued in the direction of Gueudecourt but his engine was now battering itself to pieces and he was forced to abandon the attempt. Under failing power, he managed to withdraw through Flers before breaking down completely.

There was no need for myths, for the reality of the battle was stark enough. Because of some confusion in their arrangements, the infantry did not advance from Flers to Gueudecourt as intended. Two tanks, *D5* (Lieutenant Blowers) and Legge's *D6*, therefore went on without them and found themselves duelling with two artillery batteries on the outskirts of the hamlet. Blowers inflicted some casualties on the gunners and drove them to seek cover. He waited some time for the infantry to appear, then, finding himself entirely alone, he decided to withdraw. Seeing this, the Germans returned to their guns and scored a hit that set the vehicle ablaze. Legge managed to destroy one gun and inflict further casualties before he too was knocked out. These two tanks advanced further than any others on 15 September and, had they received the intended support, it is possible that they might have achieved a breakthrough. The Germans apparently thought so too, for during the night they reinforced their artillery on this sector.

The following morning two more tanks, *D9 Dolly* (Lieutenant Victor Huffam) and *D14* (Lieutenant Cort), renewed the attack on Gueudecourt amid intense shelling – so heavy, in fact, that the infantry were again unable to advance. Both tanks had reached the edge of the hamlet when *D9*'s driver was blinded by flying splinters. With difficulty Huffam replaced him with Corporal Sanders, one of the gearsmen. *D9* began moving forward again, conforming with *D14*, which could be seen to the right, almost smothered in bursting shells. Suddenly Cort's tank lurched to a standstill, belched black smoke for several seconds, then exploded. Grinding slowly on, Huffam halted straddling a trench and opened his hatch to see that his starboard gunners were doing terrible execution among its occupants. Aware that his port gunners had stopped firing, he turned to find that both were dead and that there were several holes in the side armour. At that moment the vehicle was struck by a shell. Huffam remembered the explosion and fire breaking out, but nothing more until he came to lying across Sanders on the ground; on reflection in later years, he believed that the only possible explanation was that they had both been blown through the hatch. Many of his teeth had been smashed and his eyebrows and hair had been burned off, while Sanders had sustained ghastly leg injuries that exposed the shin bones. Huffam gave him morphia tablets, patched him up with field dressings, and sent back the other survivors, all of whom were injured or burned, hoping that they would obtain help from the infantry. No one arrived so at length he tied his belt through Sanders' and dragged the now unconscious corporal painfully from shellhole to shellhole across no-man's-land to safety.

Gueudecourt was finally captured on 26 September, mainly because of the action of a single female tank, *D4*, commanded by Second Lieutenant C. E. Storey. The previous day the infantry of the 21st Division, part of XV Corps, had been held up by machine-guns and uncut wire in front of the Gird Trench, which by now had begun to acquire a sinister reputation. Storey passed through Flers and upon reaching the trench turned parallel to it and drove from west to east, crushing the wire while his gunners fired belt after belt at the occupants. Some of the Germans fled while others dived for cover. This proved to be elusive, for at that moment a British aircraft flew the length of the trench, raking it from end to end with machine-gun fire. The infantry following up Storey's attack experienced no difficulty in taking a mile-long stretch of trench at the trivial cost of only five casualties, then went on to capture the hamlet itself, where no resistance was offered. German losses included a large number of killed and wounded, plus 370 prisoners. Despite the fact that all but two of his crew had sustained wounds of varying severity, Storey had accompanied the renewed advance until his petrol supply was exhausted; he was awarded the DSO, a remarkable honour for so junior an officer.

Both sides had already begun to assess the impact of the tanks on the land battle. For the British it could not be denied that there had been disappointments, that tank losses had been heavy and that the proportion of casualties among the crews had been higher than expected. Against this, as Haig himself put it to Swinton as early as 17 September, 'wherever the tanks advanced we took our objectives, and where they did not advance we failed to take our objectives'. Furthermore, whenever an objective was taken, infantry losses had been greatly reduced. Haig further requested the War Office to approve the building of 1000 tanks with improved armour, plus an additional 100 Mark Is. It was as a result of this rapid expansion that the companies of the Heavy Branch Machine Gun Corps grew into battalions which, starting in January 1917, were formed into brigades. Likewise, the need for a greatly enlarged training area led to the Heavy Branch leaving its interim home at Thetford in Norfolk for the rolling heathland of Bovington, Dorset, in November 1916.

Across the trench lines the initial reaction of the Germans was exactly what had been hoped for. During the late 1960s I met Werner Scholtz, a kindly man well known to collectors of those German military miniatures known as Zinnfiguren. Scholtz, a former infantryman, had witnessed one of the first tank attacks at first hand and he recalled that what had unnerved him and his comrades most was the manner in which the strange vehicles crawled towards them very slowly, lurching inexorably forward over the cratered ground. As the tanks opened fire SOS rockets were sent soaring skywards to summon artillery assistance but the gunners, used to firing barrages and concentrations, were unable to register hits unless a tank bogged down or halted. The monsters crushed their way through the entanglement and closed up to the forward trench, often dragging hundreds of yards of barbed wire behind them, which in itself was very frightening. Behind, the British infantry could be seen coming forward with bayonets fixed, cheering and yelling. Many of Scholtz's comrades were shot down by the tanks' weapons; a few, with suicidal bravery, flung grenades at the vehicles and tried to fire through the vision slits or clamber aboard across the trail wheels, but none survived; others stood frozen in shocked terror; most, recognising the futility of remaining, bolted along the communication trenches to the rear.

Prisoners described the use of such weapons as 'butchery' and the German press spoke of the tanks as 'devil's coaches'. Nevertheless, it has to be remembered that comparatively few tanks were employed and that while these achieved some success, they were only seen in ones and twos by those Germans on the spot. Thus, while panics did take place, they were purely local and were not communicated beyond the immediate scene of the action.

The attitude of the German authorities to this disclosure of a British secret weapon was mixed. At first there was some alarm and a fruitless search was made for a Mr Steiner, the Holt Caterpillar Company's representative, who had tried in vain to interest the General Staff in his machines four years earlier. Some months later, an examination of two captured tanks not only revealed several areas of mechanical unreliability, but also that the armour could be penetrated by K ammunition, which was already in general issue for use against the fire slits of concrete pillboxes, and by light trench mortars firing at low trajectory. The conclusion of the examining staff was that although the idea was interesting enough, it had a purely local application and as such represented a waste of resources. There was, therefore, no need for the German Army to develop similar vehicles of its own. For the moment, all that was necessary was for every infantryman and machine-gunner to be issued with an appropriate supply of K ammunition; for the artillery to sharpen up the response time of its heavy batteries and form direct fire anti-tank batteries positioned immediately behind the front line; and for some consideration to be given to the widening of trenches. Such an analysis was good in the short term only, for it failed to take into account the probability that the British, in the light of their own experience, would cure the mechanical faults and that the next generation of tanks would carry armour that was proof against the K round. Haig's decision to commit the tanks prematurely thus produced an unexpected dividend in that the Germans continued to underestimate the potential of the weapon until it was too late.

Blood on the Tracks

The next major engagement in which tanks were employed was at Arras in April 1917, in which 60 Mark Is and IIs took part. The weather conditions, a mixture of heavy rain and snow, were atrocious and the going was further damaged by the heavy preliminary bombardment. The sort of problem this created for the crews is vividly described by a former tank officer, Major Clough Williams-Ellis, in his book The Tank Corps:

'There [i.e., near the railway crossing at Achicourt], sunk and wallowing in a bog of black mud, were some half-dozen tanks – tanks that by then should have been miles ahead and getting into their battle position for the attack at dawn. Instead, here were the machines upon which so much depended, lying helpless and silent at all sorts of ominous angles, and turned this way and that in their vain struggles to churn their way out of the morass. About them were great weals and hummocks of mud and ragged holes brimming with black slime. The crews, sweating and filthy, were staggering about trying to help their machines out by digging away the soil under their bellies and by thrusting planks and brushwood under their tracks. Now and again an engine would be started up and some half-submerged tank would heave its bulk up and out in unsteady floundering fashion, little by little and in wrenching jerks as the engine was raced and the clutch released. Then the tracks would of a sudden cease biting and would rattle round ineffectively, the ground would give way afresh on one side, and the tank would slowly heave over and settle down again with a perilous list, the black water awash in her lower sponson.'

Needless to say, the majority of the tanks bogged down and most of the few that got forward were knocked out; one at least, having been temporarily abandoned by its crew because of mechanical problems, was destroyed by the British artillery to prevent it falling into enemy hands. There were, surprisingly, a number of successes, the most important of which was achieved, significantly, because of alterations in the original plan. Six tanks from C Battalion had been detailed to lead the 37th Division's attack on the village of Monchy-le-Preux. Of these two broke down and one ditched, but the remaining three passed through the infantry to reach the start line, 800 yards short of the objective, just before dawn. Neither they nor the infantry had been informed that the artillery had postponed their covering barrage for two hours. As the light grew and nothing happened, the tank commanders, realising that their vehicles were now silhouetted against the snow like sitting ducks, decided to begin the attack on their own initiative. The going, not having been churned up by shellfire, was reasonable and they were able to break into the village and subdue the opposition until the infantry came up to take possession. Two cavalry brigades rode forward to exploit the breakthrough but, presenting a massed target as they did, they became the focus of concentrated artillery fire and were cut to pieces. Nevertheless, the infantry were delighted to have secured a heavily fortified objective at little cost and their corps commander, Lieutenant General Aylmer Haldane, commented: 'I certainly never again want to be without tanks, when so well commanded and led.'

April 1917 also saw tanks being used on other sectors of the front, and indeed in another theatre of war. At the suggestion of an artillery officer, Colonel J. B. Estienne, the French had also begun to build chars d'assaut (assault vehicles) the previous year. The first models to appear, the Schneider and the St Chamond, were little more than armoured boxes fitted to a Holt tractor chassis, the former being armed with a 75mm gun and two machine-guns mounted in sponsons, and the latter with a forward-firing 75mm gun and four machine-guns disposed around the hull. Estienne, now a major general commanding the Artillerie d'Assaut, as the newly formed arm of service was then known, appreciated that both designs were cumbrous and lacked the obstacle crossing performance of their British counterparts. In collaboration with the Renault organisation he designed a much smaller, faster, two-man char légère that could be put into mass production. This vehicle, subsequently known as the Renault FT 17, marks a major milestone in fighting vehicle design since it was the first tank to mount a fully rotating turret, armed with either a 37mm gun or a machine-gun. On the other hand, its obstacle crossing capability was clearly limited, although a partial remedy for this was found by fitting a skid extension to the tail. The Renault FT 17 went into production shortly after it had completed its successful trials in March 1917 and by the time the war ended almost 3000 had been built.

The first French tank attack, launched on 16 April 1917 at Berry-au-Bac in Champagne, employed no less than 128 Schneiders but the result was disappointing in the extreme, 76 tanks and one quarter of the 720 crewmen being lost for little or no return. Instead of deploying before crossing their own trenches the French had opted to advance in column over comparatively few trench crossing causeways constructed by the infantry, then deploy in no-man's-land. Inevitably, the delays and bunching around the crossing sites had provided the German artillery with ideal targets and it was in those areas that most of the tank losses occurred. Again, the infantry, weary and decimated by earlier fighting, failed to keep up with the tanks with the result that when the latter took an objective they were unable to hold on to it. Three weeks later, having absorbed some of the lessons, the French mounted another attack, this time on the Chemin des Dames sector, using Schneiders and St Chamonds, and achieved better results at far lower cost. It was, in fact, on the Chemin des Dames that French armour was to achieve its first tangible success in October when, attacking on a comparatively narrow front, it played a major part in capturing the entire ridge.

April also saw tanks being used for the first time in a desert environment. The British failure to capture Gaza the previous month had resulted in the front lapsing into static trench warfare. To resolve the impasse eight tanks crewed by former members of the Heavy Branch's E Company, under the command of Major E. Nutt, were shipped to Egypt. Contrary to Nutt's advice, local commanders insisted on using them in ones and twos, so that while several individual vehicles did well during the Second Battle of Gaza no breakthrough was achieved and the attack was decisively repulsed. In November, its losses having been made good by the arrival of several Mark IVs, the Gaza Tank Detachment was wisely used to support the attack of a single infantry division and overran the Turkish defences, thereby restoring a more fluid form of warfare. After this, the Detachment played no further part in the campaign, which was conducted in the Judaean hills where the going was unsuitable.

If the Tank Corps, as the Heavy Branch had become on 28 July, thought the difficulties under which it had laboured at Arras were severe, far worse was to follow. Large numbers of the newly arrived Mark IVs were committed to the Third Battle of Ypres, sometimes known as Passchendaele, lasting from 31 July to 6 November. The Ypres Salient had a clay subsoil that retained water, and the impact of millions of shells over the previous three years had completely destroyed the drainage system. Consequently, when the weather again turned against the British at the very beginning of the offensive, torrential rain reduced the battlefield to a quagmire in which wounded men and horses simply vanished from view, while new Mark IV tanks sank into the slime to the level of their roofs. Tank officers had always carried out a detailed reconnaissance of the routes forward and it now became more necessary than ever for them to guide their vehicles into action on foot, often with fatal results. In this manner the Corps gained its first Victoria Cross, awarded to Captain Clement Robertson, whose citation reads:

'From 30 September to 4 October this officer worked without a break under heavy fire preparing a route for his tanks to go into action against Reutel. He finished late on the night of 3 October and at once led his tanks up to the point for the attack. He brought them safely up by 3am on 4 October, and at 6am led them into action. The ground was very bad and broken up by shell fire and the road demolished for about 500 yards. Captain Robertson, knowing the risk of the tanks missing their way, continued to lead them on foot. In addition to the heavy shell fire, intense machine-gun and rifle fire was directed at the tanks. Although knowing that his action would inevitably cost him his life, Captain Robertson deliberately continued to lead his tanks when well ahead of our own infantry, guiding them carefully and patiently towards their objective. Just as they reached the road he was killed by a bullet through the head, but his objective had been reached, and the tanks in consequence were enabled to fight a very successful action.

'By his very gallant devotion, Captain Robertson deliberately sacrificed his life to make certain the success of his tanks.'

Three further awards of the Victoria Cross were made to members of the Tank Corps during the war, respectively to Captain R. W. L. Wain, Lieutenant C. H. Sewell and Lieutenant Colonel R. A. West; all, like Robertson's, were made for dismounted actions and all were posthumous.

So rare were tank successes during Third Ypres, and so heavy the cost in bogged and shattered vehicles, that a general feeling grew throughout the Tank Corps that it had been deliberately put in a situation where it would be discredited. That was not altogether true, although the Corps did have its opponents. What was true was that the infantry commanders expected the tanks to produce decisive results, and this they were unable to do. Therefore, reasoned the infantrymen, they were an extravagance and their crews would be better employed as replacements for the infantry battalions, which by now were becoming seriously short of men. Both Elles, now a brigadier general, and Fuller, his Chief of Staff, were aware of the feeling and conscious that unless the Corps was given a chance to prove itself over ground of its own choosing the pressure for its reduction would grow. Haig, though he still believed that tanks were simply a tool of trench warfare subordinate to the infantry and artillery, was not unsympathetic and gave permission for the Tank Corps staff to plan such an operation.

The area selected was the firm, rolling, chalk downland of the Cambrai sector, which drained well and still possessed a surface that had not been cut up by incessant shellfire. Fuller proposed a sort of 'tank raid' against the enemy's communications in Cambrai, employing the entire Corps along a north-easterly axis on the comparatively narrow frontage bounded by the Canal de l'Escaut to the east and the Canal du Nord to the west. The idea met with the approval of General Sir Julian Byng, commander of the Third Army, in whose area this sector lay, but he insisted that the raid should become a major offensive. Elles and Fuller had serious reservations about this as the reserves normally required to sustain such an offensive had all been sucked into the fighting in the Ypres Salient, but they desperately needed Byng's support and had perforce to agree.

The first factor to be considered in the planning of Operation GY, as the offensive was to be known, was the nature of the enemy's Hindenburg Line. The first of the defences consisted of a series linked outposts and behind these were a number of strongpoints ahead of the forward trench. The forward trench itself contained numerous dugouts, reinforced concrete observation and machine-gun posts with steel cupolas and had been dug to a width of not less twelve feet, rendering it tank-proof as far as the Germans were concerned. Some 300 yards to the rear was the similarly constructed support trench. Each of the three trenches was fronted by wire entanglements some 50 yards wide. The sector was considered to be a quiet one in which troops could be rested after taking their turn in the Ypres Salient. Their experience there, and against French tanks earlier in the year, had engendered such confidence among the Germans in their anti-tank measures that they had begun thinning out their direct fire artillery batteries. During the third week of November interrogation of prisoners suggested that the British were up to something and various plans were prepared to meet a conventional attack; none of these allowed for the possibility of an attack by tanks.

The Tank Corps' answer to the widened trenches was to employ huge brushwood fascines carried on the tanks' roofs. A drill was worked out in which tanks were to work in sections of three in arrowhead formation. Having crossed the outpost line, the leading tank would turn left on reaching the forward trench and rake it with its starboard guns; the second tank would drop its fascine into the main trench, then cross and turn left to rake both it and the support trench; the third tank would drop its fascine into the support trench, then cross and turn left to rake the defenders from the rear. Finally, the first tank would come forward with the infantry and, after the trenches had been cleaned out, tanks and infantry would rally on the objective and resume the advance. This drill, it will be noted, left each tank section with one fascine in hand to meet unforeseen contingencies. Interdependence was to be the keynote of the attack; if machine-guns held up the infantry, the tanks would deal with them; if the tanks were halted by field artillery, the infantry would be responsible for disposing of the gunners. To ensure the closest possible cooperation the infantry were to advance in files 25 yards behind the tanks, thereby enabling them to pass quickly through the gaps crushed in the enemy wire.

It had been the Tank Corps' intention to smash through the enemy front in a series of concentrated blows, but once again its staff had been forced to give way to the demands of the infantry commanders who wanted the tanks spread along the front so that the ultimate allocation of tank battalions was as follows:

III Corps (Right Flank)
12th Division – C and F Battalions
20th Division – A (less one company) and I Battalions
6th Division – B and H Battalions
29th Division (in immediate reserve) – one company A Battalion
IV Corps (Left Flank)
51st (Highland) Division – D and E (less one company) Battalions
62nd Division – G Battalion plus one company of E Battalion

Altogether, the Tank Corps fielded a total of 474 tanks for the battle. Of these, 376 were Mark IV gun tanks while the remainder, mainly thin skinned Mark Is and IIs converted for other duties, included 32 fitted with grapnels for towing wire out of the path of the Cavalry Corps, who it was hoped would exploit the breakthrough, 54 supply tanks and a handful of bridging and communications vehicles. After dark on 19 November this huge assembly of armoured vehicles moved slowly along taped lanes up to their attack positions, the noise of their engines masked by sustained machine-gun fire. This considerable administrative feat was all the more remarkable in that it was achieved without the enemy suspecting anything. Elles, knowing how much depended on the success of the operation, had already drafted and distributed his famous Special Order No 6 to tank commanders:

'1. Tomorrow the Tank Corps will have the chance for which it has been waiting for many months – to operate on good going in the van of the battle.
2. All that hard work and ingenuity can achieve has been done in the way of preparation.
3. It remains for unit commanders and for tank crews to complete the work by judgement and pluck in the battle itself.
4. In the light of past experience I leave the good name of the Corps with great confidence in their hands.
5. I propose leading the attack of the centre division.

Hugh Elles, B. G.
Commanding Tank Corps
19th November 1917'

Zero hour was 0620 the following morning. Some five minutes earlier Elles, puffing his pipe, strolled into H Battalion's area, in the approximate centre of the six-mile line of attack. The men could see that the general was carrying an ashplant walking stick, although there was nothing unusual in that; during Third Ypres most officers had taken to carrying one while undertaking route reconnaissances for their vehicles across the unspeakable terrain, using one or two hands to push the stick into the ground and so determine the approximate weight it would tolerate. There was, however, something different about Elles' stick, which seemed to be wrapped in coloured cloth. After talking briefly to one of the company commanders he indicated a tank named *Hilda*, commanded by Lieutenant T. H. de B. Leach and, clambering aboard, he shut the sponson door behind him.

Almost at once there was a tremendous crash as 1000 guns opened fire simultaneously and the tanks began rolling forward. The supporting barrage was deliberately timed to coincide with the tanks' advance so that the going would not be too severely damaged, and it included a high proportion of smoke shells to blind the enemy's artillery observers. Aboard *Hilda*, Elles asked the driver to accelerate a little until the tank was in full view of the entire battalion, then opened the top hatch and stuck his head and shoulders out, unfurling the cloth around his stick as he held it aloft. The wind caught it and revealed it as a brown, red and green tricolour flag – the Tank Corps colours symbolising the unofficial motto, 'From Mud, Through Blood, To The Green Fields Beyond'; and thus a determination to put an end to the horrors of trench warfare. It had been many years since officers of such seniority had personally led their troops into battle and Elles' decision to do so was inspirational; later, Fuller was to comment, 'It was spiritually the making of the Tank Corps, and in value it transcended all our work.'

After a while Elles dropped down into the interior of the vehicle. Quite properly, he did not attempt to influence the way Leach fought his vehicle and spent most of the time peering over the driver's shoulder. *Hilda* crossed the outpost line at the second attempt, but at 0815 became ditched in the Hindenburg Line's forward trench. As he was no longer able to influence the battle directly, Elles decided to return to his headquarters on foot. Major Gerald Huntbach, Leach's company commander, recalled him waving cheerily as he walked briskly by, pipe aglow, with the now bullet-ripped Corps flag in his hand; following behind, at a respectful distance, came several groups of shaken German prisoners.

That there is a great similarity in many survivors' accounts of Cambrai is due to their experiences being much the same, at least during the first day's fighting. They describe the thunder of the supporting barrage; the growing smokescreen rising above the enemy trenches with SOS rockets and flares bright against it; the strengthening grey light suddenly warmed by sunrise; the ground mist through which the tanks rolled forward flat out, achieving almost four miles per hour over incredibly firm going; crushing gaps through the wire and crawling across the enemy trenches on their huge fascines; and Germans everywhere, fleeing, surrendering or fighting back as best they could. Perhaps the most graphic of these accounts is that of Captain D. G. Browne of G Battalion, recorded in his book *The Tank in Action*, published in 1920 when his memories were still vivid:

'The immediate onset of the tanks inevitably was overwhelming. The German outposts, dazed or annihilated by the sudden deluge of shells, were overrun in an instant. The triple belts of wire were crossed as if they had been beds of nettles, and 350 pathways were sheered through them for the infantry. The defenders of the front trench, scrambling out of dugouts and shelters to meet the crash and flame of the barrage, saw the leading tanks almost upon them, their appearance made the more grotesque and terrifying by the huge black bundles they carried on their cabs. As these tanks swung left-handed and fired down into the trench, others, also surmounted by these appalling objects, appeared in multitudes behind them out of the mist. It is small wonder that the front Hindenburg Line, that fabulous excavation which was to be the bulwark of Germany, gave little trouble. The great fascines were loosed and rolled over the parapet to the trench floor; and down the whole line tanks were dipping and rearing up and clawing their way into the almost unravaged country beyond. The defenders of the line were running panic stricken, casting away arms and equipment. The Hindenburg Reserve, with its own massive entanglements, went the way of the first trenches; and so far our following infantry had found little to do beyond firing on the fugitives and rounding up gangs of half stupefied prisoners. It was now broad day, the mist was thinning, and everywhere from Havrincourt to Banteux on the canal was rout and consternation.'

To some extent, the Germans contributed to their own misfortune. Their standing operational orders insisted that lost ground must be recovered immediately by counter-attack and, amid the mist and growing confusion several units were committed without full understanding of what was involved; the tanks caught them deployed on open ground and they were cut to ribbons. In the midst of their other troubles came the spine-chilling discovery that the K ammunition was little use against the Mark IV.

Yet, despite the overwhelming success of the British assault, the Germans did not give the tanks a free run. During the fight for Havrincourt village they climbed on the roofs of several G Battalion tanks, puncturing and setting fire to the spare petrol tins carried there, so forcing the crews into the open. On the left flank *Early Bird*, belonging to the detached company of E Battalion, was confronted by a hundred or so of the enemy coming forward with their hands raised; some 50 yards short of the tank the Germans scattered, revealing several machine-guns that opened fire at once. However, guns blazing, *Early Bird* charged them down and something akin to a massacre took place. But elsewhere, having overcome its initial shock, the German artillery was beginning to take its toll.

The worst check of the day was sustained on the sector where D and E Battalions were leading the advance of Major General G. M. Harper's crack 51st (Highland) Division. Harper was fiercely proud of his division's reputation and careful with the lives of his men, who in return were staunchly loyal to him and referred to him unofficially as Uncle. The problem was that Harper's ideas on war were so reactionary that they bordered on the Crimean. At one period he had strenuously opposed the introduction of the machine-gun, which he argued would lower the overall standard of musketry. He regarded the whole idea of the Cambrai offensive as 'a fantastic and most unmilitary scheme', and, for no intelligible reason, he altered the tanks' trench clearing drill beyond recognition. He further ordered his troops to keep a minimum of 100 yards behind the tanks, on the grounds that the heavy fire that would inevitably be directed at the latter would cause casualties among those nearby; that this would also deprive the tanks of infantry support if they encountered anti-tank artillery he considered to be of lesser importance. Tank Corps officers who complained to Third Army Headquarters on both counts were simply told that Harper was master in his own house.

Nevertheless, by 0820, 51st Division had taken its first objectives without the slightest difficulty. Then, although neighbouring divisions continued to advance, Harper decided to stick to the timetable and allowed his men a lengthy two-hour rest. During that time a Major Krebs, commander of the 27th Reserve Jäger Regiment, had organised the defences of Flesquières village, lying in the division's path beyond the shallow ridge of the same name. In addition, the guns of the 108th and 282nd Field Artillery Regiments, both of which had been trained in anti-tank techniques, were dragged from the camouflaged gun pits that had previously concealed them from air reconnaissance, and positioned on the reverse slopes of the ridge.

52

It was 0930 before Harper gave the order for the advance to be resumed. By the time the tanks were crawling across the crest of Flesquières Ridge the gap between them and the infantry had widened to 400 yards, and suddenly they were confronted by lines of field guns spitting fire. A brief but murderous duel ensued in which sixteen tanks, hit time and again, were turned into blazing wrecks from which no survivors emerged. Other tanks worked their way onto the flanks of the German batteries and shot down the gunners. However, the Germans also had their heroes that day. One gun, believed to have accounted for five tanks, remained in action to the bitter end, served by a single gunner who continued to load and fire until he too fell behind the gunshield.

Their ranks thinned, D and E Battalions pressed on. As they passed Flesquières village, consisting of well constructed houses with cellars and a château surrounded by a wall, Krebs and his men went to ground. They surfaced again as soon as the Highlanders appeared with the result that for the rest of the day Harper's division became embroiled in a protracted struggle for the village. The commander of the neighbouring 62nd Division offered to assist by attacking Flesquieres from the rear, but Harper declined, feeling that this would involve loss of face. The result was that Krebs and his 600 defenders retained possession well into the night, then withdrew of their own accord.

Elsewhere, despite this setback, the tanks had reached the Green Fields. In the distance the crews could see Cambrai while, closer to hand, there were farms, villages and woods, unspoiled by war, all lying beneath the watery light of the fading sun. A six-mile wide breakthrough had been achieved and, because of it, the church bells at home would ring in celebration of a great victory for the first time since the war began.

At that moment the frantic German High Command was ordering formations to converge on Cambrai from all over the Western Front; some would reach the area during the night, others next day, but few believed that they would arrive in time to stop the British exploiting their victory and breaking out into open country. Yet, to their bewildered relief, no such exploitation took place. The responsibility for this must rest with the Cavalry Corps, the headquarters of which now lay miles to the rear of the forward troops. There, staff officers had simply refused to believe the speed with which the breakthrough had been achieved, and the commanders of formations which had followed up the advance were reluctant to proceed without orders from above. By the time Cavalry Corps Headquarters had accepted the reality of the situation daylight was fading and the horses, which had been standing to all day, needed watering. The only worthwhile contribution was made by one squadron of the Canadian Fort Garry Horse which, tired of the constant debate and endless inaction, took matters into its own hands and overran a German battery, to be all but destroyed in turn when they ran into a superior force of infantry. Guderian's comment that 'five cavalry divisions had been unable to rip through a thin screen of a few machine-guns and rifles' is, therefore, something of an oversimplification, even if the result was the same.

To put the Tank Corps' achievement on 20 November into perspective, it had punched out a salient six miles long and four miles deep in the German line in a matter of twelve hours at a total cost of 4000 casualties, including 648 of its own; during a conventional offensive such as Third Ypres three months had been needed to make similar gains, at an approximate cost of 250,000 casualties. Tank losses during the day were 65 written off or seriously damaged by enemy action, 71 broken down with mechanical problems and 43 ditched or out of action for other reasons. The Germans sustained 10,500 casualties, including a high proportion of prisoners, and the equipment captured included 123 guns, 79 trench mortars, 281 machine-guns and large stocks of ammunition and stores.

After the first day of Cambrai the conduct of war could never be the same again. It is worth recalling the reply of Winston Churchill, who had been so closely involved with the introduction of armoured vehicles into the British armed services, when in later years he was asked what else the generals could have done other than wage attritional trench warfare:

'I answer, pointing to the Battle of Cambrai, this could have been done. This in many variants, this in larger and better forms, ought to have been done.'

For a few days after the great tank attack the Third Army continued to maintain pressure on the Cambrai sector. However, as comparatively few tanks were available to support the dog-tired infantry, and the German defence was becoming stiffer by the hour, there was little further progress. Operation GY wound down and the tanks began withdrawing to their railhead. Then, on 30 November, the Germans launched a counter-offensive which astonished everyone with its speed and weight. By rounding up 63 assorted tanks and crews and sending them back into battle this was finally brought to a standstill on 7 December, but by then the Germans had recovered much of the ground they had lost, and a little more besides, as well as inflicting comparable casualties to those they sustained on 20 November.

Unknown to the British, the Germans were using a recently developed offensive technique of their own, first employed with spectacular success against the Russians at Riga in September and then against the Italians at Caporetto the following month. In essence, this contained three elements. First, there was a brief but extremely heavy and carefully orchestrated artillery programme designed to neutralise the opposing artillery, destroy or isolate specific strongpoints along the front with a mixture of high explosive, gas and smoke shells, and provide the assault infantry with a rolling barrage. Second, the leading wave of the assault consisted of storm troops, whose instructions were to bypass centres of resistance, leaving them to be dealt with by the follow-up waves, and press on at speed into the enemy artillery and administrative areas. Third, the attack was supported continuously by waves of ground-attack aircraft. Had this technique been developed earlier, it is possible that it could have won the war for Germany. As it was, General Erich Ludendorff, the German Army's First Quartermaster General and de facto Commander in Chief, intended using it to obtain decisive results in the west the following spring, employing specially raised and trained storm troop battalions and the numerous formations released from the eastern front by the collapse of Russia.

By now the German Army had changed its mind about tanks, although it can hardly be said to have embraced the subject with full-blooded enthusiasm. In late 1916 a department had been set up to design a German tank. A Holt tractor was obtained from Austria and a lengthened version of this, powered by two Daimler 100hp engines, served as the basis for the vehicle. A large, top-heavy, box-like superstructure was fitted, consisting of 30mm armour plate at the front and 15–20mm plate elsewhere. Armament consisted of one Russian 57mm gun in the front plate, two machine-guns on each side of the hull and two more at the rear. No less than sixteen men were stuffed into the restricted space available yet, despite the obvious discomforts involved, they enjoyed a smoother ride than their British counterparts because of the sprung suspension. On good going the tank could achieve 5mph, although its obstacle crossing capability did not compare with that of British machines. Known as the A7V (the abbreviated title of the design committee), it was accepted for service after being demonstrated to the General Staff at Mainz on 14 May 1917, and 100 were ordered. However, delivery of these was so slow that, while seven five-tank operational units had been formed by December, only three were equipped with the A7V; the remainder were issued with cannibalised British Mark IVs which had been battle-damaged and were awaiting recovery when the German counter-offensive overran part of the Cambrai battlefield. These units were scheduled to take part in the series of offensives Ludendorff planned for the spring, with the very limited role of subduing predesignated strongpoints.

Over the winter the Tank Corps also received new vehicles, exchanging its Mark IVs for the much improved Mark V, some mention of which has already been made. Two battalions, the 3rd (formerly C) and the 6th (formerly F) were re-equipped with a new three-man tank, officially known as the Medium A, but more commonly as the Whippet. The design of this differed from that of earlier British tanks in that it employed a forward engine and transmission compartment, housing twin Tylor 45hp motor bus engines, separated from the fixed crew compartment, armed with four Hotchkiss machine-guns, at the rear. Although capable of 8.3mph, the complexity of the controls meant that the driver had to acquire something of the skills of a church organist if he was not to stall one or both engines while making a change of direction. In theory, the Whippet was intended to cooperate with the cavalry and so avoid the sort of hiatus that had taken place at Cambrai. In practical terms, the two arms were incompatible, for when there was no opposition the horsemen would canter ahead, leaving the tanks far behind, until they were brought to a standstill by a single machine-gun; then the tanks would have to catch up and take the lead, alone and unsupported by the infantry, until the problem was resolved. Nevertheless, despite so essentially flawed a theory, the Whippet battalions were to give a remarkably good account of themselves.

Ludendorff commenced his series of offensives on 21 March and they continued until the middle of June, striking different sectors of the Allied line and advancing over thirty miles in places. There were times when the Germans appeared to be on the verge of a clean breakthrough, but somehow they were held, often by scratch units consisting of clerks, batmen, cooks and drivers. During this period the Tank Corps provided Lewis gun teams to reinforce the hard-pressed infantry, and performed a role described by Elles as 'Savage Rabbits', in which the tanks lay hidden until the moment came for them to launch sudden counter-attacks on the unsuspecting enemy. Such tactics, though they inflicted serious casualties, were quite unsuited to the heavy Mark IVs and Vs which, having launched their local attacks, often found that the storm troopers had worked round their flanks and were continuing to advance faster than the tanks could withdraw to the most recently established defence line. The cost in broken down tanks, blown up or set ablaze to prevent their falling into enemy hands, was heavy; as was the loss of crews cut off and captured or killed. These considerations did not apply to the much faster Whippets, which first went into action with the 3rd Battalion on 26 March to close a four-mile gap that had been torn in the Third Army's front.

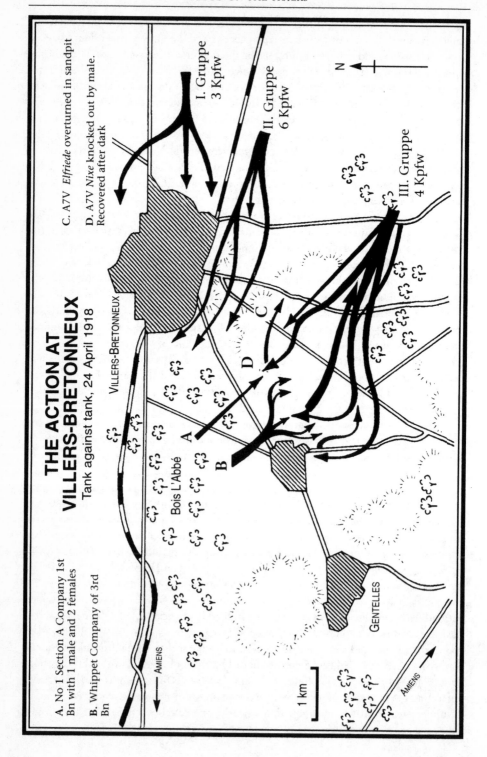

THE ACTION AT
VILLERS-BRETONNEUX

Tank against tank, 24 April 1918

A. No 1 Section A Company 1st
Bn with 1 male and 2 females

B. Whippet Company of 3rd
Bn

C. A7V *Elfriede* overturned in sandpit

D. A7V *Nixe* knocked out by male.
Recovered after dark

I. Gruppe
3 Kpfw

II. Gruppe
6 Kpfw

III. Gruppe
4 Kpfw

VILLERS-BRETONNEUX

Bois L'Abbé

GENTELLES

AMIENS

AMIENS

1 km

N

'Twelve proceeded towards Colincamps, at which village the leading company arrived just as the Germans were entering it. It is reported that the Germans mistook this new pattern tank for their own, and began to cheer them. They were quickly disillusioned, however, as the tanks, proceeding at a rapid pace, first mopped up a strong patrol, then retook the village, and then went on towards Serre, where they dispersed three bodies of troops, each above a battalion strong. All twelve Whippets returned safely after their successful enterprise, which appears to have been particularly opportune, preventing as it did a hostile encircling movement.'

The Germans had, in fact, used four A7Vs and five captured Mark IVs on the St Quentin sector five days earlier, but this went unrecorded by the British, partly because the attack had been delivered in thick mist, and partly because the defenders were either dead or had been overrun and captured. It was inevitable that, sooner or later, a clash would occur between the two opposing tank arms and this took place at Villers-Bretonneux, east of Amiens, on 24 April. Here, in the Bois l'Abbé, were the three tanks of No 1 Section, A Company, 1st Battalion, their crews recovering from a rain of gas shells with which the Germans had opened a fresh attack. The section, commanded by Captain J. C. Brown, consisted of two female and one male tank, the last under the command of Second Lieutenant Francis Mitchell. The enemy had already penetrated Villers-Bretonneux when Brown received orders to proceed to the disconnected series of trenches known as the Cachy Switch Line, to the south of the village, and hold it at all costs.

The tanks began moving forward at 0845. Brown travelled with Mitchell until, nearing the forward trenches, he was warned by the infantry that there were German tanks about and dismounted, intending to control the movements of his section on foot, as was the custom. Mitchell, whose crew were working shorthanded because of gas casualties, has left us with two written accounts of the subsequent engagement, of which the following is the more graphic.

'I informed the crew and a great thrill ran through us all. Opening a loophole, I looked out. There, some 300 yards away, a round, squat-looking monster was advancing; behind it came waves of infantry, and further away to the left and right crawled two more of these armoured tortoises. So we had met our rivals at last! For the first time in history tank was encountering tank!

'The six-pounder gunners crouched on the floor, their backs against the engine cover, and loaded their guns expectantly. We still kept on a zigzag course, threading gaps between the lines of hastily dug trenches, and coming near the small protecting belt of wire we turned left. The right gunner, peering through his narrow slit, made a sighting shot and the shell burst some distance beyond the enemy tank. No reply came. A second shot boomed out, landing just to the right, but again there was no reply. More shots followed. Suddenly a hurricane of hail pattered against our steel wall, filling the interior with myriads of sparks and flying splinters. Something rattled against the steel helmet of the driver sitting next to me and my face was stung with minute fragments of steel. The crew flung themselves flat on the floor. The driver ducked his head and drove straight on. Above the roar of our engine sounded the staccato rat-tat-tat-tat of machine-guns and other furious jets of bullets sprayed our steel side, the splinters clanging against the engine cover. The Jerry tank had treated us to a broadside of armour piercing bullets.

'Taking advantage of a dip in the ground.....we manoeuvred to get the left gunner onto the moving target. Owing to our gas casualties the gunner was working single-handed, and his right eye being swollen with gas, he aimed with the left. Moreover, as the ground was heavily scarred with shell holes, we kept going up and down like a ship in a heavy sea, which made accurate shooting difficult. His first shot fell some fifteen yards in front, the next went beyond, and then I saw the shells bursting all round the tank. He fired shot after shot in rapid succession every time it came into view.

'Nearing the village of Cachy, I noticed to my astonishment that the two females were slowly limping away to the rear. Almost immediately on their arrival they had both been hit by shells which tore great holes in their sides, leaving them defenceless against machine-gun bullets, and as their Lewis guns were useless against the heavy armour plate of the enemy they could do nothing but withdraw.

'Now the battle was left to us, with our infantry in their trenches tensely watching the duel like spectators in the pit of a theatre. As we turned and twisted to dodge the enemy's shells I looked down to find that we were going straight into a trench full of British soldiers, who were huddled together and yelling at the tops of their voices to attract our attention. A quick signal to the gearsman seated at the rear of the tank and we turned swiftly, avoiding catastrophe by a second.

'Then came our first casualty. Another raking broadside from the German tank and the rear Lewis gunner was wounded in both legs by an armour piercing bullet which tore through our steel plate. We had no time to put on more than a temporary dressing and he lay on the floor, bleeding and groaning, while the six-pounder boomed over his head and the empty shell cases clattered all round him..... We turned again and proceeded at a slower pace. The left gunner, registering carefully, began to hit the ground right in front of the Jerry tank. I took a risk and stopped the tank for a moment. The pause was justified; a well-aimed shot hit the enemy's conning tower, bringing him to a standstill. Another roar and yet another white puff at the front of the tank denoted a second hit! Peering with swollen eyes through his narrow slit, the gunner shouted words of triumph that were drowned by the roar of the engine. Then once more he aimed with great deliberation and hit for the third time. Through a loophole I saw the tank heel over to one side; then a door opened and out ran the crew. We had knocked the monster out! Quickly I signalled to the machine-gunner and he poured volley after volley into the retreating figures.

'My nearest enemy being now out of action, I turned to look at the other two, who were coming forward very slowly, while our six-pounder gunners spread havoc in the ranks of the advancing German infantry with round after round of case shot. [This actually consisted of iron balls strung at intervals on steel wire; the effect can be imagined - *Author*] Now, I thought, we shan't last very long. The two great tanks were creeping relentlessly forward; if they both concentrated their fire on us we would be finished. We fired rapidly at the nearest tank, and to my intense joy and amazement I saw it slowly back away. Its companion also did not appear to relish a fight, for it turned and followed its mate and in a few minutes they had both disappeared, leaving our tank the sole possessor of the battlefield.'

Shortly after this, Mitchell's tank was disabled by a direct hit from an artillery shell and he and his crew spent the next two hours in an infantry trench. Both Mitchell and Brown, who had continued to direct the action on foot after the two females had retired, were awarded the Military Cross. Mitchell's reference to a gearsman indicates that his own tank was a Mark IV and confirms that the 1st Battalion had not yet been fully re-equipped with Mark Vs. One interesting aspect of the engagement is that, due to their inexperience, the German tank commanders failed to identify No 1 Section's male as the priority target and instead concentrated their fire on the two females, which could do them no harm at all.

That same morning, and just a few hundred yards away, a British reconnaissance aircraft spotted two storm troop battalions resting in a hollow near Cachy. The pilot recrossed his own lines and dropped a message to a 3rd Battalion Whippet company, lying up some distance to the rear, suggesting that if the tanks hurried they would catch the enemy in the open. The company commander, Captain Thomas Price, seized the opportunity at once. In line abreast, some 40 yards apart, his seven Whippets charged into the hollow from the north, taking the storm troopers completely unawares. Those who sought cover from the rattling machine-guns in shell holes were pursued and crushed. At the end of their run the tanks turned and combed the hollow again, their tracks now slimy with blood and human remains. Both German battalions were completely dispersed with a loss in killed and wounded amounting to over 400 men. One Whippet, which ignored Price's instructions to avoid showing itself on the skyline, was knocked out, and three were damaged – at the time it was thought by artillery, although it was subsequently confirmed that the four had come within range of a solitary A7V that had remained in the area.

The German version of the tank versus tank fighting at Villers-Bretonneux, as related by Guderian, indicates that four A7Vs were present. The loss of one, overturned in a sandpit, is acknowledged. It is also acknowledged that the tank engaged by Mitchell was hit several times and was abandoned by the crew, who lost five of their number killed; later, they returned and drove it back to their own lines. This sheds considerable doubt on the previously accepted British version that the overturned tank recovered from the sandpit, with the assistance of French engineers, was the A7V engaged by Mitchell. Indeed, it seems highly probable that the overturned tank, named *Elfriede*, was already lying in the sandpit when Brown's section reached the battlefield, since it had come to grief at about 0845, just as the British Mark IVs were moving off. Furthermore, one of Mitchell's very specific accounts says that his opponent 'tilted slightly,' while that quoted above used the words 'heel over'; had the tank crashed onto its beam ends he would undoubtedly have said so. The roots of this misunderstanding almost certainly lay in the fact that these two A7Vs were advancing along closely parallel axes, though perhaps not at the same time; that Mitchell's opponent was recovered after dark; and that when the stranded *Elfriede* was found lying in her sandpit it was assumed, naturally enough, that it was she who had fallen victim to Mitchell's gunnery.

According to the German account, which there is no reason to doubt, the commanders of the two remaining A7Vs were unaware of the presence of the Mark IV male, let alone that a tank duel was in progress. Having completed their task, which was to eliminate British machine-gun positions at Cachy, they began to retire towards their assembly area. Last to go was Lieutenant Bitter's tank, on the extreme left of the German line. While turning, he was horrified to observe Price's murderous counter-attack on the

German infantry. He engaged the Whippets at ranges between 200 and 700 yards with his 57mm gun, but after four rounds the weapon's firing-pin spring broke and he was forced to continue with his machine-guns. His claim to have destroyed three Whippets was, as we have seen, optimistic. After Price's company had retired Bitter's tank became the rallying point for the survivors of the two infantry battalions, with whom he remained until 1445 that afternoon, some 400 yards short of Cachy, which had been the day's objective.

In May and June Ludendorff shifted the weight of his offensives to the French sector of the front. Unlike the British, the French used their tanks in a series of local counter-attacks, some of which were more successful than others. By the middle of July it was apparent that the German Army had shot its bolt and the initiative was once more returning to the Allies. On 18 July French and American divisions mounted a Cambrai-style counter-offensive into the flank of a salient which had developed south of Soissons, using 211 Schneiders and St Chamonds and 135 Renault FT 17s. On the first day a four-mile penetration was achieved but thereafter the pace of the advance slowed and, instead of being trapped in the salient as the French intended, the Germans were able to conduct an orderly withdrawal, albeit at the cost of 25,000 men taken prisoner.

The British, too, had begun mounting local counter-attacks, the most successful of which captured the village of Hamel, located on a shallow ridge to the north of Villers-Bretonneux, of 4 July. Here, 60 Mark Vs, manned by crews from the 8th and 13th Battalions, supported the attack of ten Australian battalions and four American infantry companies on a frontage of three and a half miles. The objective, one and a half miles within enemy lines, was secured within an hour, and with it 1500 prisoners, two field guns and 171 machine-guns. The Australians sustained 775 casualties and the Americans 134; Tank Corps losses amounted to thirteen wounded but none killed. Five tanks damaged by shellfire were recovered during the next two nights.

On 23 July the Tank Corps' 9th Battalion, on loan to General Debeney's neighbouring First French Army, led an attack which enabled the infantry to take their objectives, capturing 1858 prisoners, five field guns and 275 machine-guns at a cost of 1891 casualties of its own. The 9th Battalion lost 54 men killed or wounded, plus eleven of its 35 tanks knocked out. The French were so impressed by the 9th's efficiency that Debeney issued a Special Order of the Day acknowledging its contribution, awarded the unit a corporate Croix de Guerre, and requested that it should wear the badge of the French 3rd Division, with which it had fought.

The success of operations such as these encouraged the Allies to approve the suggestion of General Sir Henry Rawlinson, commander of the Fourth Army, for a major offensive, spearheaded by tanks, to be mounted east of Amiens using the British III Corps, the Australian Corps and the Canadian Corps. The choice of the Dominion troops was intentional as their divisions were still largely up to strength and had retained their aggressive edge, while most British formations were still recovering from their ordeals of the spring; infantry brigades, for example, had already been reduced from four to three battalions, within which tired rifle companies were operating at about half their established strength.

Preparations for this, the greatest armoured assault of the war, were cloaked in secrecy and covered by an elaborate deception plan. There were similarities to Cambrai, but in the eight and a half months which had intervened better tanks had been issued and experience had produced better battle drills. There was also to be greatly increased artillery and air support, and adequate reserves were available. The only ominous aspect of the plan was that, once again, the Cavalry Corps had been given the task of exploiting the breakthrough.

With exception of the 1st Tank Brigade (7th, 11th and 12th Battalions), which was re-equipping and training with Mark Vs, and the 9th Battalion, placed in General Reserve and refitting after its recent action with the French, the entire Tank Corps had been assembled for the battle. The 4th Tank Brigade (1st, 4th, 5th and 14th Battalions) was allocated to the Canadian Corps, on the right of the attack; the 5th Tank Brigade (2nd, 8th, 13th and 15th Battalions), plus the armoured cars of the 17th Battalion, were allocated to the Australian Corps in the centre; and the 2nd Tank Brigade (reduced to the 10th Battalion to provide additional punch on other sectors) was allocated to the British III Corps, covering the left flank of the attack, north of the Somme. With the exception of the 1st and 15th, which had been issued with the Mark V*, all these battalions were equipped with Mark Vs. Finally, the 3rd Tank Brigade (3rd and 6th Battalions with their Whippets) was allocated to the Cavalry Corps. Altogether, the attack would involve 324 heavy tanks, with a further 42 in immediate reserve, 96 Whippets, 120 supply tanks, over half of which were allocated to the infantry, and 22 gun carriers, which were cut-down Mark I chassis intended to ferry 60-pounder or six-inch howitzers across no-man's-land, although they were more commonly used to transport ammunition and supplies.

Before coming to the battle itself it is necessary to examine the state of the German Army at this period. The storm troop battalions had creamed off the infantry's best soldiers, most of whom had been killed in the spring offensives. Inevitably, those now serving in the line regiments were of a different quality and, while far from mutinous, they had become increasingly disillusioned and war weary. The victory they had been promised had simply not materialised, despite the sacrifices they had willingly made. Now, thrown once more onto the defensive, they saw divisions being broken up to provide replacements, of which were never enough to fill the gaps in the ranks; yet, across the lines, more and more American divisions were reaching the front, and the Allies had begun using their tanks with an ever-increasing frequency.

It was the tanks which troubled the average German infantryman most, for he had few defences against them now that their thicker armour had rendered his K ammunition all but useless. True, a new anti-tank rifle, the Tankgewehr 18, a bipod-mounted brute of a weapon which at 200 yards was capable of punching a 13mm round through 20mm armour plate, thicker than anything the British possessed, had been issued, but there were never enough to go round. True, also, a 37mm anti-tank gun for use in the forward trenches was under development, although none had reached the front as yet. In the final analysis, the German infantryman could rely only on the artillery for support, as he always had done.

The defence of the Amiens sector was the responsibility of the German Second Army, commanded by General von der Marwitz. There were seven divisions in the line and four more in reserve, but their average strength amounted to barely 3000 men and only two of them were considered to be fully battleworthy; nor were their defences anything like the formidable obstacles that had been so easily overcome at Cambrai. During the fortnight prior to the attack Second Army Headquarters had received so many groundless tank alerts from the nervous tenants of the trenches that it not only became indifferent to them but also issued a sharp criticism of the troops for crying wolf.

There was some justice in this, for there had been no tank activity on the days in question. In fact, the tank trains did not arrive until the last possible moment, and the muted rumble of hundreds of fighting vehicles moving forward under cover of darkness from railheads into operational assembly areas was deliberately drowned by the drone of low-flying aircraft. At 0420 on 8 August the guns crashed out, firing a Cambrai-style rolling barrage while the long lines of tanks and infantry moved forward through a dense morning mist, thickened to the consistency of a fog by smoke shells. In places it was impossible to maintain precise direction but as long as a tank commander conformed to the tidal flow of men and machines he could not go far wrong.

For the Germans, the mist and smoke added a nightmare dimension to the experience. Above the thunder of the artillery was the rising growl of many engines and constant clattering of tracks, all hidden from view beyond the grey curtain. Then the

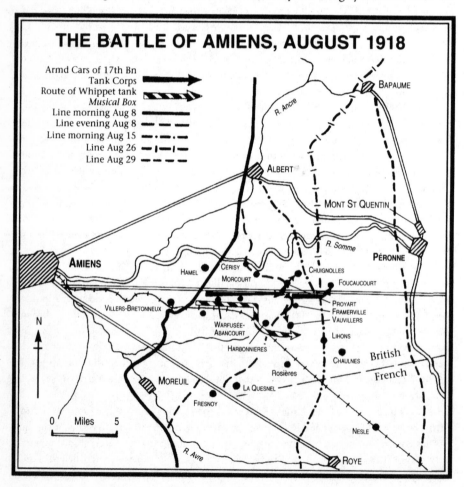

THE BATTLE OF AMIENS, AUGUST 1918

Armd Cars of 17th Bn Tank Corps
Route of Whippet tank *Musical Box*
Line morning Aug 8
Line evening Aug 8
Line morning Aug 15
Line Aug 26
Line Aug 29

huge shapes appeared, indistinct at first but suddenly in focus, groups of them stretching away to the right and left, splintering their way through the wire entanglements, spitting fire, rearing and crushing. The German machine-gunners, tough and dedicated to the bitter end, fired belt after belt to no avail until they were shot down or vanished beneath the churning tracks, but most of their comrades, lacking any means with which to defend themselves, surrendered in large numbers rather than be raked by the tanks' machine-guns or be torn apart by the terrible case shot.

The first objective was reached between 0630 and 0700 and it was now that the tanks entered their hardest fight of the day. At about 0645 the mist began shredding away, exposing the tanks to the German field gun crews. A series of murderous duels took place in which neither side could afford to give quarter. When it was over, 109 tanks had been knocked out or disabled but the German divisional artillery regiments, whose gunners fought to the muzzle, had ceased to exist. The tide rolled on, reaching the second objective by 1100 and the third by 1300. It was now up to the Cavalry Corps to exploit the gap, eleven miles wide and up to seven miles deep, that existed in the German line.

Wishing to avoid the mistakes that had been made at Cambrai, the Corps Commander had already authorised his divisional commanders to press on well beyond the third objective, although some were hesitant about doing so without specific orders. Lieutenant J. B. Robertson, commanding one of the 14th Battalion's tanks, describes the brave sight presented by the 3rd Cavalry Division as it moved forward over ground captured by the Canadian Corps:

'Streaming up the long southern track they came, headed by a regiment of lancers. As far as the eye could reach there were trotting columns of horses while in the middle track batteries of horse and field artillery were arriving at a gallop. A cloud of dust on the northern track heralded the Whippets, 40 of them, moving almost as fast as the artillery and going hell for leather for the next objective. The whole spectacle was one which none of us had ever expected to see in France and one we would never forget.'

Exactly the same words were used by a German officer to describe the result of a mounted attack without tank support:

'The cavalry pushed forward, and in the next moment became a chaotic tangle of horses, splattered in their own blood, staggering on broken legs, or galloping riderless through our infantry.'

The incompatibility of cavalry and tanks, already noted above, was a sore point with Tank Corps Headquarters which believed that if the Whippet battalions had been left under its control they would by now be operating between five and ten miles ahead, turning the enemy's defeat into a complete rout. As it was, a few individual Whippets managed to break free of the cavalry's restraining hand to cause mayhem in the German rear areas, and the story of the most famous of these forays, that of Lieutenant C. B. Arnold's *Musical Box*, forms part of Chapter 14; the only real exploitation of the day, carried out by the armoured cars of the 17th Battalion, has already been described in Chapter 2.

Nevertheless, Amiens was a stunning victory, won at very modest cost by the standards of earlier battles, and it was the Tank Corps' crowning achievement of the war. Of it, the German report has this to say:

'When darkness descended on the battlefield on 8 August, the heaviest defeat suffered by the German Army since the beginning of the war had become an accomplished fact. The total losses of units employed on the Second Army sector can be put down as from 650 to 700 officers and 26,000 to 27,000 men. More than 400 guns and an enormous quantity of machine-guns, mortars and other war material were lost. More than two-thirds of the total German losses were due to prisoners. Almost everywhere it is evident that German soldiers had surrendered to the enemy or thrown away rifles and equipment, abandoned trench mortars, machine-guns and guns, and sought safety in flight.'

Ludendorff described 8 August as 'The Black Day of the German Army,' not so much because of the physical damage it had sustained, but rather because the mass surrenders indicated shattered morale and 'put the decline of our fighting powers beyond doubt.' He informed the Kaiser that a German victory was no longer possible and the latter instructed his Secretary of State to initiate peace negotiations through the offices of the neutral Queen of the Netherlands. In the meantime the Army would continue to fight defensively in the hope that the gains made in the early months of the war would place Germany in a strong negotiating position.

After Amiens the Allies initiated a general advance along the Western Front and the character of the war once again became mobile. On the British sector, where the heaviest fighting took place, the Tank Corps, joined by its 16th Battalion and the American 301st Tank Battalion fighting in British vehicles, took part in every major battle, fighting its way across the old Somme battlefield and again breaking the Hindenburg Line, this time using wooden cribs instead of cumbrous fascines. It was never able to field as many tanks as it had at Amiens, and at one period was forced to resurrect some of the recently discarded Mark IVs to keep its battalions viable. The Whippet battalions, released from cavalry control, were often in the van of the action.

On 8 October, appropriately enough during the series of operations known as the Second Battle of Cambrai, the first battle between tank units was fought east of the old battlefield and south of the town. Here the 12th Battalion, seriously under strength and still equipped with the Mark IVs which it had hoped to discard, was supporting the continued advance of Byng's Third Army. On the right B Company, under Major H. S. Inglis, had only seven tanks with which to support the advance of IV and V Corps up the valley of the Torrent d'Esnes to the higher ground near Esnes itself; in the centre, Major D. H. Richardson's C Company, with eight tanks, was to capture Seranvillers and the large farms surrounding the village jointly with VI Corps; on the left, the five tanks of A Company, commanded by Major H. Campbell, were to take Niergnies in conjunction with XVII Corps. Unknown to the British, the remnants of the German tank detachments, amounting to one A7V and ten captured Mark IVs, had been concentrated near Niergnies and were simultaneously spearheading a counter-attack.

The Twelfth's A Company began moving slowly forward at 0430. It was still dark and their already limited vision was further curtailed when the supporting artillery began dropping some of its smoke shells short. At the level crossing on the Cambrai-Rumilly road *L8*, commanded by Lieutenant H. Carmichael, became ditched in the cellar of a house. Carmichael dismounted to see what could be done and was immediately surrounded and taken prisoner. Observing that the attitude of his captors was less than half-hearted, he quickly talked them round and they became prisoners in their turn, being handed over to the British infantry when it arrived.

Meanwhile, the company's remaining four tanks had continued to advance on the objective, clearing pockets of resistance as they did so. After Niergnies had been taken the infantry consolidated on the rising ground beyond. The light was growing stronger and, as the tanks prepared to withdraw to their rally point, the strength of the German artillery fire began to increase. At that precise moment *L16*, commanded by Captain Roe, was stationary at Mont Neuve Farm, to the east of Niergnies. Suddenly, Roe spotted a number of Mark IVs crawling towards him from the east. At first, he took them to be C Company tanks, although they were approaching from the wrong direction. Not until the gap had closed to 50 yards did black crosses on the sides of the new arrivals become visible, and then recognition was mutual. Roe immediately opened fire with one of his six-pounders, causing the leading enemy tank to lurch to a standstill; simultaneously, *L16* was penetrated twice, one round smashing its way through the forward visor, wounding Roe and killing his driver. The engine stopped and the crew bailed out. Roe ran to the nearest British tank, Second Lieutenant Warsap's *L9*, only to find that this, too, had sustained critical damage. Five of the crew, including the six-pounder gunners, had been wounded, and the rest were fighting a fire. Nevertheless, Warsap continued to engage the enemy with a Lewis gun until the vehicle sustained a further hit, causing the fire to break out afresh and forcing the crew into the open. The third A Company tank, *L12*, was penetrated twice and knocked out before the crew realised that the strange Mark IVs were hostile. Only one tank, commanded by Second Lieutenant A. R. Martell, remained, but this too was in difficulty with a leaking radiator which had caused the engine to overheat and cut out, while enemy fire had put three of the Lewis guns out of action. Martell ordered the crew out and under cover then joined an artillery officer in swinging round a captured 77mm field gun to knock out a second enemy vehicle.

The infantry on both sides had understandably gone to ground while the armoured juggernauts fought it out. Now, with A Company eliminated at the cost of only two of the enemy's tanks, it seemed that the first round had undoubtedly gone to the Germans, who despatched two of their females to pin down the British infantry pulling back from the crest towards the village. However, the battle was far from over and in a very short space of time the situation was to be transformed.

To the south C Company had seen the VI Corps infantry onto their objectives and was also preparing to retire to the rally point when, on the extreme left *L54*, a female commanded by Second Lieutenant J. B. Walters, encountered two enemy females which had just emerged from the fight with A Company. In theory neither side could damage the other, but Walters engaged them both with continuous machine-gun fire, expending several thousand rounds. After a while it seems as though German crews had had enough of bullet splash and flying shards of metal, for both tanks ceased firing and made off towards their own lines; on the way, Walters noted with satisfaction, one was hit and destroyed by artillery fire.

Nearby, two more hostile females had attacked the infantry east of Seranvillers, forcing them to pull back. The intruders were promptly pounced upon by two C Company males, *L45* and *L49*, commanded respectively by Second Lieutenants Clark and Sherratt, and penetrated repeatedly with six-pounder fire until they began to burn and were abandoned by their crews. With this the heart seemed to go out of the enemy counter-attack, enabling the infantry to move forward again and reoccupy the positions they had been forced to vacate. That afternoon the Germans abandoned Cambrai, having sustained the loss of 10,000 prisoners and 200 guns during the day's fighting. 'The operations,' wrote the 12th Battalion's historian, 'Provided the first tank action over unspoiled, open country, such as may be seen anywhere in England and north Europe. The tank was constructed originally to surmount trenches, craters and similar obstacles; and in those early days there was a feeling that when it came to open warfare again the infantry could manage adequately by themselves; but in fact it was found that the further we advanced, and the more the conditions approximated to the old idea of open warfare, the more essential the tank had become.' And, that being the case, it was inevitable that in future wars, tank would be pitted against tank as a matter of course.

The last tank action of the war took place near the Forest of Mormal on 5 November when eight Whippets of the 6th Battalion supported an attack by the 3rd Guards Brigade. During the previous three months the Tank Corps had been in action more or less continuously at one point or another along the British line. In the period between tanks first being committed to battle at Delville Wood and the Armistice it had sustained the loss of 879 killed, 5302 wounded and 935 missing. By 1918 the Corps had grown from a handful of adventurous spirits to a strength of approximately 18,000. Of them, Winston Churchill wrote to Prime Minister Lloyd George: 'it is no exaggeration to say that the lives they have saved and the prisoners they have taken have made these 18,000 men the most profit-bearing we have in the Army.'

By now the Allies had become firmly convinced of the value of the tank, which would have played the major role on the battlefield if the war had continued. All their tank corps would have been dramatically expanded and re-equipped. Faster designs than the Whippet were already well advanced and the British and Americans were cooperating on a much improved heavy tank, the Mark VIII or International, which was to have been built in large numbers.

Fuller had already projected an operational method for the coming year. During the first phase of this he envisaged a 'disorganising force', consisting of fast tanks with air cover, effecting a penetration and driving deep into the enemy's rear areas to destroy senior headquarters, while bombers and ground attack aircraft multiplied the confusion with attacks on communications centres and supply dumps. As soon as the enemy's command structure had been eliminated, a 'breaking force' of heavy tanks, infantry and artillery would shatter the enemy's front. After this a 'pursuing force' consisting of more fast tanks, lorry-borne infantry and cavalry would harry the defeated army until its will to fight had been destroyed.

Marshal Foch, now Allied Supreme Commander, accepted Fuller's proposals and would have implemented them had the need arisen. Now generally referred to as Plan 1919, they formed the keystone of the Blitzkrieg technique and have been in dominant element in military thought ever since.

The Gembloux Gap and Flavion
The First Tank Battles of World War II

In the years that followed World War I the tank quickly acquired the basic layout which it has retained to this day, namely a driving compartment, a fighting compartment surmounted by a fully traversable turret housing the main armament, and an engine compartment. All first class armies produced several classes of tank including heavy or infantry tanks intended to support infantry operations, medium or cruiser tanks for use in armoured formations, and light tanks for reconnaissance and imperial policing; for a while, there was also a fashion for 'tankettes', which were little more than tracked machine-gun carriers, but these had little practical application and by 1939 only the Italian and Japanese armies continued to use them in any numbers.

The great debate of the inter-war years centred around the correct manner in which tanks were to be employed. All armies, save one, earmarked a proportion of their tank strength for infantry support, but as to the rest the debate continued with only untested theories as a guide, and the only conflict in which tanks were used in any numbers, the Spanish Civil War, failed to produce clear lessons.

The United Kingdom's contribution was largely intellectual. The ideas of Fuller, Martel and, later, Liddell Hart, regarding autonomous armoured formations in which the movements of the tanks, together with that of their organic and fully mechanised infantry, artillery, anti-tank and engineer units, could be controlled through a radio net, were proved to be viable by the Royal Tank Corps in a series of experimental exercises. The climate of the times, however, was very similar to that of today, in that political and economic expediency took priority over military necessity, and there was a chronic shortage of modern equipment. Thus, although the Royal Armoured Corps was formed from the Royal Tank Corps and the mechanised cavalry regiments in 1939, the outbreak of war found the Army with only two embryonic armoured divisions, the 1st in England and the 7th in Egypt, and one embryonic tank brigade, all of them seriously under equipped.

The real beneficiary of these ideas was the German Army, which had begun forming panzer divisions as soon as Hitler repudiated the restrictive clauses of the Treaty of Versailles in 1935. Having analysed in minute detail the reasons for their defeat in 1918, the Germans were in a better position than most to appreciate the potential of the tank. The more progressive elements within the Army, supported by Hitler, accepted the British theory that in striking deep to achieve the strategic paralysis of the enemy, movement actually provided its own defence, but modified this to conform with their own traditional concept of the Annihilation Battle (Vernichtungsgedanke) in which one or both of the enemy flanks were turned in upon the centre, which was then crushed, just as the Roman army had been at Cannae. Where the German Army took the theory a stage further, however, was in forming its panzer divisions into panzer corps.

The German armoured corps, in fact, had not expected to go to war as early as 1939 and its equipment programme lagged far behind the Führer's political ambitions. Its light tanks, the PzKw I and II, were armed only with machine-guns and were of little use in the tank battle; the PzKw III medium, intended to carry the burden of the tank versus tank fighting, was armed with a 37mm gun but was in very short supply; and the PzKw IV, equipping the fourth or heavy company of each panzer battalion, was armed with a short 75mm howitzer that was effective against anti-tank guns but not armour plate. If the occupation of Czechoslovakia had not yielded a substantial dividend in the form of Czech-built PzKw 35T and PzKw 38T tanks which, because of their 37mm guns, could be substituted for the PzKw III, the Corps would have been hard pressed to meet its commitments. Again, the long-term intention had been that the panzer divisions' organic motor rifle regiments (redesignated panzer grenadiers in July 1942) would be carried in SdKfz 251 armoured half-tracks. However, in May 1940, on the eve of the invasion of the West, only two out of 80 motor rifle battalions were so equipped, a proportion which actually fell throughout the war although the number of panzer grenadier battalions rose to its peak of 226 in September 1943. The one aspect of mechanised warfare the armoured corps steadfastly refused to consider was providing support for infantry operations. After some heated debate on the subject it was decided that this function would be carried out by the artillery, using assault guns based on a PzKw III chassis with a 75mm howitzer mounted in the front plate of a fixed superstructure. No assault gun units were available for service in 1939, but four batteries took part in the 1940 campaign in the West and thereafter the number rose steadily, with the Assault Artillery achieving the status of an élite branch of service within its own right.

The French Army, on the other hand, devoted a high proportion of its considerable tank strength to infantry support, although during the 1930s some consideration had also been given to the role of mechanised cavalry regiments. By 1939 three Divisions Légères Mécaniques (DLM) had been formed with the intention that they should perform the same roles once carried out by horsed cavalry, such as providing a screen for the army while it deployed, reconnaissance en masse and exploitation of a victory. In 1940, having taken heed of the performance of the German panzer divisions in Poland, the Army decided that it needed a heavier type of armoured formation and formed three Divisions Cuirassées (DCR). Equipped with heavy and infantry tanks, these were best suited to the breakthrough role. In practical terms, however, neither the DLM nor the DCR fulfilled the flexible concept of British and German armoured formations, nor could any real flexibility be achieved as only one tank in five was fitted with a radio. Nor was it intended that the operations of the DCR should be controlled by a corps headquarters, an omission that was to have serious consequences.

Many of the French tanks in service in 1939, particularly the Char B and the Somua, were well armed and armoured for their day. Unfortunately, the preference for one-man turrets proved to be a fatal flaw. In practice this meant that the tank commander was simultaneously required to direct his driver; choose ground for tactical advantage; load, aim and fire his guns; use the radio, if fitted; and, if he was an officer, try to control the other tanks under his command. To ignore any of these functions was to court disaster, yet such was their combined pressure that none could be adequately performed. Significantly, with the exception of their smallest tanks, the British and Germans rejected the concept of the one-man turret in the correct belief that a well-drilled two- or three-man turret crew was essential to the fighting efficiency of the vehicle.

The course of the 1940 campaign in western Europe is too well known to require more than the briefest summary here. On 10 May von Bock's Army Group B invaded Holland and Belgium. In response, the British Expeditionary Force and the best of the French armies advanced north into Belgium in the belief that the Germans were repeating the Schlieffen Plan of 1914. While these movements were taking place, von Rundstedt's Army Group A, spearheaded by the bulk of the German armour, advanced through the Ardennes, which the French had mistakenly believed to be tank-proof. After crossings of the Meuse had been secured, three panzer corps broke through the French Second and Ninth Armies and carved a corridor 40 miles wide across northern France, reaching the Channel coast on 20 May. They then swung north into the rear of Franco-British-Belgian forces in Belgium, which were now trapped inside a pocket. Holland had already surrendered and when Belgium capitulated on 27 May, placing the British and French troops within the pocket in an untenable position, the Dunkirk evacuation followed. Regrouping rapidly, the Germans renewed their offensive against the remaining French armies, now deployed along a line stretching from the Somme to the Aisne and the Maginot Line, on 5 June. The French resisted fiercely, but the best of their armoured formations had already been lost and the Luftwaffe controlled the skies. A breakthrough was inevitable and, to prevent the whole of France being overrun, an armistice was requested, and granted, on 25 June.

The speed with which these events developed has sometimes been allowed to obscure the fact that several large scale tank battles did take place between French and German armoured formations, and that these, excluding the occasional tank versus tank encounters during the Spanish Civil War, were the first clashes of their kind between the armoured corps of first class armies; nor, indeed, were they quite so one-sided as the above might suggest.

Of the two panzer corps serving with von Bock's Army Group B one, Schmidt's XXXIXth, with only 9th Panzer Division under command, was fully engaged in driving across Holland to the relief of the German parachute and air-landing troops who had formed an airborne carpet between Moerdijk and Rotterdam, effectively wrecking the Dutch defensive strategy. The second, Hoepner's XVIth, containing the 3rd and 4th Panzer Divisions, crossed the Maastricht Appendix and, bypassing the formidable defences of Fort Eben, which had fallen to a daring glider assault during the first hours of the war, drove on into central Belgium. Nominally, the two-regiment panzer brigade of each division contained 140 PzKw Is, 110 PzKw IIs, 50 PzKw IIIs and 24 PzKw IVs. Far behind the probing armoured cars of the divisional reconnaissance battalions, the tanks moved in a keil or wedge, with the rest of the division following. Little resistance was encountered until, on 12 May, they entered the Gembloux Gap, a stretch of open country lying between the Sambre and the Dyle rivers, to the east of the old battlefields of Ligny and Quatre Bras. Here, in the area of Hannut, they found themselves confronted by Prioux's Cavalry Corps, screening the French First Army as it arrived from the south to occupy positions along the Dyle, and a major clash of armour became inevitable.

Prioux's Corps, consisting of the 2nd and 3rd DLMs, was considered to be an élite and it had not been affected by the malaise that was eating away at some sections of the French Army. Each division contained an armoured brigade, a reconnaissance brigade with armoured car and mechanised infantry units, an artillery regiment and an engineer battalion. The overall tank strength of the Corps amounted to 174 Somuas (47mm gun and 40mm armour) and 87 Hotchkiss H.35s (37mm gun and 34mm armour), plus 40 AMR light tanks. Against the heavier French machines, Hoepner could only pit his 100 PzKw IIIs and 48 PzKw IVs, the 30mm armour of which compared unfavourably with that of their opponents.

Nevertheless, while Prioux's tanks, engaged in their screening role, were spread across a wide area, Hoepner's panzer regiments each shook out into two parallel battalion groups with their PzKw IIIs in front, the lighter PzKw Is and IIs on the flanks and the PzKw IVs ready to provide support with their howitzers as required. Simultaneously, the German anti-tank units established fronts through which their tanks could retire if worsted, while the motor rifle regiments occupied the most import terrain features within their divisional areas.

Neither side had fought a major tank battle before, although the Germans were the more experienced, having played a prominent part in the Polish campaign the previous year. The impression left by various accounts of the fighting is one of unfamiliarity, with engagements taking place at closer range than was necessary and little of the use of ground for protection that would become instinctive in subsequent engagements. As the first flat traces of armour piercing shot began criss-crossing the battlefield the supporting artillery of both sides also opened fire and shellbursts began to fountain among the tanks. At first it began to seem as though the Germans, outnumbered but fighting in more concentrated formations, were having the worst of the exchange. However, the Luftwaffe, a tactical air force, had already attained air superiority and its Ju 87 dive-bomber squadrons, summoned by the panzer divisions' forward air controllers, soon intervened, howling down on the ranks of the French to drop their bombs with deadly accuracy. This provided an additional and unnerving distraction for the already overworked French tank commanders in their one-man turrets. The Germans began to notice that their opponents moved little and tended to fight their own individual battles with little mutual cooperation. While their practised turret crews maintained a steady rate of fire, unit commanders used their superior radio net to concentrate their tanks against individual French vehicles. By the time that dusk had put an end to the day's fighting the battlefield was littered with burning tanks, but both sides were still holding their positions. The course of the battle, and its sequel the following day, is described by Lieutenant Colonel Heinrich Eberbach, commander of Panzer Regiment 35, which, with Panzer Regiment 36, formed the armoured brigade of 4th Panzer Division.

'The 2nd Battalion was ordered forward to Hannut, to secure the important road junction there. In Villers-le-Peuplier it wiped out an enemy column approaching from the south and took many prisoners. On leaving the village to continue the advance in a south-westerly direction Second Lieutenant Rauschenbach's tank was hit by an anti-tank gun and he was wounded. Many of the prisoners taken were French and gradually it became clear that south of Hannut the enemy was present in strength.'

'During the night tank leaguers were protected by our riflemen's outposts. On 13 May the whole panzer brigade renewed the battle. We moved forward at 1245, to be struck by fire from Merdorp. French tanks could be seen clearly through our binoculars. A violent fire fight began, with no tactical pattern. Our aim was simply to bypass Merdorp, leaving the tanks in the village to be dealt with by the infantry following behind. In rolling past the enemy flank the 1st Battalion hit eight tanks, but more heavy fire was directed at us by enemy tanks from the water-tower hill at Jandrain. In the teeth of this we struggled to reach the low hills between Jandrain and Jandrenouille. It had become a real witches' cauldron. Our ammunition began to run short and the light platoons were sent back to bring up more. A number of our own tanks had been hit and lay halted and silent, with dead and wounded aboard. The situation could hardly be described as rosy. The enemy was almost invisible, but we did not give up. To the west our heavier tanks had formed a firing line and the 3rd Company was engaged in clearing Jandrain. The crew of a French Somua, positioned in a hollow, gave up the fight and surrendered. In Jandrain the 3rd Company took 400 prisoners and captured four anti-tank guns and five tanks. At length the ammunition arrived and we continued our advance in the direction of Ramillies-Offus. On the way we passed eight knocked-out French tanks, testimony to the fierce battle we had just fought.'

Each side claimed to have knocked out about 100 of the other's tanks in the battle, although these figures may be somewhat exaggerated. However, when Prioux, having successfully covered the First Army's deployment, withdrew from the field, he left his damaged vehicles behind, whereas Hoepner was able to recover and repair a proportion of his vehicle casualties. The Germans, also, could claim to have succeeded in their object, for the tank battle had served to concentrate French attention on the Low Countries rather than Army Group A's advance through the Ardennes to the Meuse, which was where the real threat lay. As a direct result of the battle General Maurice Gamelin, the Allied Commander in Chief, detached the 1st and 2nd DCRs from his central reserve and sent them north to Belgium.

By the evening of 12 May von Rundstedt's three panzer corps had reached the Meuse. The following day bridgeheads were secured and, working under sustained air attack, engineers put in the bridges that would allow the tanks to cross. On the 14th, Gamelin realised the enormity of the mistake he had made and ordered all three DCRs to converge on the German bridgeheads and wipe them out: the 1st would be brought down from central Belgium and attack from the north; the 2nd, still in transit, would be halted and come in from the west; the 3rd, lying close to Sedan, would attack from the south. Significantly, as each DCR arrived it would come under the command of the local infantry corps commander, so that rather than taking the form of a coordinated counter-stroke controlled by a single headquarters, the entry of the most powerful element in the French armoured corps was reduced to the level of a series of local attacks. Furthermore, Gamelin neither attuned to the pace at which the war was being conducted, nor aware of the effects that order followed by counter-order would have on the DCRs themselves.

Thus, while 1st DCR was alerted during the morning, its commander, instead of being briefed by radio, was told to report to the headquarters of XI Corps, some twenty miles distant, at 1330; yet, when he arrived he received the simple order to attack towards Dinant as soon as possible. By the time he had returned and issued orders to his unit commanders most of the day had been wasted. When, at last, the Division's 68 Char Bs and 90 Hotchkiss H.35s began moving towards the front, they did so along roads choked with refugees and fugitives from the French Ninth Army, which had been routed on the Meuse. Painfully slow progress resulted in such heavy petrol consumption that the time of attack was postponed from dawn until noon on 15 May so that refuelling could take place.

At first light, however, Hoth's XV Panzer Corps broke out of its bridgehead. In the lead was the 7th Panzer Division, reinforced by the leading armoured regiment of 5th Panzer Division, much of which had been delayed by congested roads beyond the Meuse. This meant that the most important elements of 7th Panzer's tank strength, its 132 PzKw 38Ts and 36 PzKw IVs, was supplemented by some 20 PzKw IIIs and another dozen PzKw IVs, plus light tanks in proportion, producing an extremely powerful formation. In command was Major General Erwin Rommel, a dynamic, thrusting officer whose tireless energy had won him Imperial Germany's highest decoration, the Pour le Mérite, in World War I. It was typical of his style of command that during the approach march to the Meuse he had ordered his tanks to fill up at civilian pumps rather than wait for their own supply echelons.

Near Flavion, Rommel's leading elements ran into 1st DCR, replenishing in close leaguer. Because of this, only a proportion of the latter's Char Bs, referred to by the Germans as Kolosse, were able to get into action at once, but those that did made a formidable impression. Their armament, consisting of a 47mm gun in the turret and a 75mm howitzer in the front plate, dealt telling blows and their 60mm armour was impervious to the German 37mm guns. On the other hand, their radiator louvres, located in the side of the hull, were vulnerable, as were their tracks, and, picking these as their aiming points, the German gunners slowly began to score kills. Rommel, however, recognised that in its present semi-mobile condition 1st DCR did not present a threat and, calling up his dive-bombers to batter the French formation he swung round its flank and began driving west into the hinterland of France. 1st DCR had still not sorted itself out when, about 1400, it was assailed by the now understrength 5th Panzer Division. The nature of the fighting is described by a junior officer serving in a Char B company.

'"En avant!" orders the *Adour*, the Captain's tank. The *Gard* is on my right, the Captain is to the right of the *Gard*. At that moment a shell strikes the armour on the left side! Towards the road, red lights flare up on the level of a low hedge; another shell in the armour plate! I hesitate to shoot back, because I thought it was a mistake by one of ours; then I traversed my turret towards the flames and shoot off five high-explosive shells at the hedge, after which nothing moves any longer. I continue my advance and arrive at the woods which mark the edge of the plateau, and it is there that the battle begins. The driver shouts: "A tank on the edge of the wood in front of us!" It was certainly an enemy! A PzKw IV on which I directed the fire of the 75mm. Near a burning German tank, men are climbing and crawling towards the undergrowth. The whole of our left flank is crowded with big German tanks; I can make them out more or less indistinctly, because they are camouflaged, broadside on and immobile.

'At this moment the co-driver of the Captain informs me that the Captain, wounded in the stomach and legs, is handing over command to me. The new PzKw IVs burst into flames under our fire, but my radiators are smashed in; my 75mm is hit on the side of the muzzle and remains in the position of maximum recoil; I continue with the 47mm. Feeling myself harassed, I try to change position and move myself to a thicket further to the south. The wood is being hammered by 105mm (artillery) and shell-holes open up not far from us. From a distance I can make out the *Gard*, the door of whose turret is open. On my right is a knocked out tank of the 28th Battalion. The line of German tanks form a semi-circle of vehicles which I estimate in number at between fifty to sixty.

'I give the order to my tanks to retire. *Ourcq* and *Yser* withdraw slowly, while I observe *Herault* burning.'[1]

As the French were pushed slowly back, a substantial number of their tanks had to be abandoned because of mechanical failure or fuel shortage and these were destroyed by their crews. When 1st DCR broke contact and withdrew at dusk, it continued to shed vehicles along its route so that by dawn on 16 May it had been reduced to a mere seventeen tanks and was no longer a force to be reckoned with. It had undoubtedly inflicted damage on both of Hoth's panzer divisions, although its claim to have knocked out 100 German tanks was almost certainly on the high side; and, again, having retained possession of the battlefield, the Germans were able to recover a proportion of their casualties.

At the time of Gamelin's redeployment, 2nd DCR's tanks were travelling north by rail and its wheeled vehicles by road. The divisional commander, unfortunately, had not been informed of the change of plan so that when he reported to First Army headquarters on 14 May he was told that his division had been reallocated to the Ninth Army. The latter, however, advised him that it was to form part of a blocking force known as Army Detachment Touchon and he therefore ordered his units to converge on Signy l'Abbaye the following. However, on 15 May Reinhardt's XLI Panzer Corps broke out of its bridgehead and, at about 1700, overran part of the divisional artillery and the rest of the wheeled units headed for safety south of the Aisne, leaving the armoured demi-brigades to the north. 2nd DCR therefore never went into action as a division, although its tanks continued to fight under local control.

3rd DCR was actually in position ready to launch its own counter-attack at 1600 on 14 May when, incredibly, the local Corps Commander cancelled the operation and broke up the tank battalions to form a twelve-mile line of static pillboxes covering the southern flank of the bridgehead. When Guderian's XIX Panzer Corps broke out to the west the following day the division was reassembled and thrown into a protracted and ultimately fruitless battle of attrition against the southern shoulder of the penetration.

On 16 May the French acknowledged that they had lost a decisive battle; the three DCRs had been squandered to no purpose and the Cavalry Corps was fully committed in Belgium. The next day, in a desperate attempt to halt the German advance the newly formed and virtually untrained 4th DCR, commanded by Major General Charles de Gaulle, was flung against the southern wall of the panzer corridor at Laon, but was beaten off. On 21 May the incomplete British 1st Army Tank Brigade delivered a counter-stroke against the corridor's northern wall at Arras, cutting Rommel's division in two for a while and savaging his motor rifle regiments before it could be halted. As a result the German drive was suspended for 24 hours, but the outcome of the campaign had already been decided.

From the planning stage onwards, the French High Command, the size of whose contribution ensured their role as decision makers among the Allies, had been

---

1 Quoted from Alistair Horne's *To Lose a Battle*.

caught wrong footed. No dishonour whatever attached to the French tank crews, who had fought courageously, hard and well despite being handicapped by dubious doctrine, flawed design and hideously bad staff work. The Germans, on the other hand, had unreservedly accepted the full potential of the iron fist and had single-mindedly created the conditions in which it could be used to smash a way through to the Channel.

Barbarossa
Raseiniai and Brody-Dubno, June 1941

'**W**e have only to kick in the front door and the whole rotten house will come tumbling down!' was Adolf Hitler's excited prediction on the eve of his invasion of the Soviet Union. Superficially, there was support for such a view, for the winning combination of his armoured troops and the Luftwaffe had won him a series of stunning victories in Poland in 1939, in the West in 1940, and in Yugoslavia and Greece in the spring of 1941; furthermore, the performance of the lumbering Red Army in Finland and Poland had been far from impressive.

It was, nevertheless, foolish to despise any intended victim of unknown potential, let alone one of such size, on the basis of this evidence. Colonel General Franz Halder, the German Army's Chief of Staff, was later to comment on the faulty premise that one could set in train such an enormous undertaking in a spirit of adventure and hope to carry it through without the industrial capacity to support troops operating in the immensity of the Russian heartland in all seasons. Hitler himself, once the enormity of his error had become apparent, declared that he never would have initiated such a war had he known that the Soviet tank production statistics contained in the then Major General Heinz Guderian's book *Achtung - Panzer!*, published in 1937, were accurate.

The greater part of the Russian heavy industry was, in fact, devoted to the production of war material and by 1941 the USSR possessed the largest tank fleet in the world, estimated at 24,000 vehicles, including amphibious tankettes, light tanks, cruiser tanks and medium and heavy infantry tanks. Many were poorly conceived and of little use in the tank battle, but in some respects Russian designers were far ahead of their Western counterparts. The KV (Klimenti Voroshilov) series of heavy tanks, produced by the Kotin team, was protected by 90mm armour, heavier than anything in service elsewhere. The T34, a product of the Koshkin team, employed the Christie high-speed suspension and possessed mathematically angled hull armour that overhung the tracks, giving ballistic protection equivalent to twice that of the 45mm plate used. Both designs were armed with a 76.2mm gun, were capable of being upgunned, were driven by the same high performance diesel engine, were fitted with wide tracks which enabled them to negotiate mud, sand and snow in a manner which the German tanks could not, and both were in quantity production when the invasion began. The T34, in fact, provided such a balanced combination of firepower, protection and mobility that it is now regarded as the starting point of modern tank design.

That the Soviet armoured corps had adopted a 76.2mm calibre as standard for its main armament at a time when the Germans were making the transition from 37mm to 50mm is a story in itself. Stalin had appointed one of his old Civil War cronies, Marshal G. I. Kulik, as his Chief of Artillery, largely because he was too dim to be devious and could therefore be relied on. Although Kulik knew very little about his profession, or anything else for that matter, he was much given to making Olympian but totally groundless pronouncements, one of which was that German tanks were being fitted with 100mm armour plate. As luck would have it, a team of Russian experts was visiting German tank production plants at the time and its members flatly refused to believe their hosts' assertion that the PzKw IV, then being fitted with 50mm frontal armour, was Germany's most recent design. The team's suspicions tended to support Kulik's assertions and as a result the Red Army's newest generation of tanks was fitted with guns capable of penetrating the thicker, if as yet imaginary, plate.

The fundamental weaknesses of the Soviet armoured corps lay not in its tanks but in its lack of a coherent doctrine, its monolithic command and control apparatus and its primitive communications. In the mid-1930s Marshal Mikhail Tukhachevsky had attempted to infuse the Red Army with modern ideas and began forming tank corps. These were unwieldy but, given time, they might well have solved most of their problems. In Stalin's eyes, however, the Army was the servant of the Party; independent thought in so sensitive an area was not to be tolerated since it generated Praetorianism and therefore posed a threat to his own position. Tukhachevsky became the first victim of the Great Purges of 1937–8 and was followed by thousands of officers who were either shot after summary trials, sent to labour camps or simply vanished without trace. Those who survived were too cowed to do anything more than obey direct orders; displays of personal initiative were not only dangerous, but were also pointless in a situation where all power was concentrated in the Centre. When General D. G. Pavlov, commander of the Soviet contingent which fought in the Spanish Civil War, returned home, his view was that large armoured formations were a waste of time and that the only correct role for the tank was infantry support. To the accompaniment of such dogmatic claptrap as 'victory resting with the bayonets of the proletariat', Tukhachevsky's corps were broken up. However, when the poor performance of Soviet armour during the Winter War with Finland and the occupation of Poland and the Baltic States was contrasted with that of the German armour in the West, the Kremlin hurriedly decided that the armoured corps should be formed again. Each of the new mechanised corps contained two tank divisions, each of two tank regiments with 200 tanks apiece, a motor rifle regiment and an artillery regiment; and a motor rifle division. Each corps received a quota of the new KVs and T34s, but most of their tank strength consisted of a hotch-potch of older vehicles, many of which were in poor mechanical condition, so that to the basic incompatibility between designs was added a spares problem of monumental proportions.

As if this was not bad enough, the poor Soviet tank officer, having been forced to change horses twice in mid-stream, was badly equipped to execute such orders as he was given. Radio links existed between corps, divisional and regimental headquarters, but only in the most fortunate formations did these extend as far down as battalion. From battalion downwards, detailed briefings were necessary before every phase of an operation. Once a tank unit was committed to action, the degree of internal control available to its commander was little better than had existed on a World War I battlefield. Nor was he permitted to vary his segment of the master plan to suit local requirements; the result was that, to the bewilderment of the Germans, failed attacks were regularly repeated over the same ground, with similar results and heavy losses.

The German armoured corps, in sharp contrast, was at the peak of its abilities. Since the campaign in the west the number of armoured divisions had been doubled by reducing the original panzer brigades to single regiments. The battles against the British Matildas and French Somuas and Char Bs had revealed that the 37mm gun could not be relied on as an armour defeating weapon, and while nothing could be done to rectify this in the case of the Czech tanks, most of the PzKw IIIs had been upgunned with a 50mm L/42 or L/60 main armament, while the thickness of frontal armour had been increased to 50mm. Of the 3332 tanks available for Operation 'Barbarossa', the codename for the invasion of the USSR, only 965 were PzKw IIIs, the balance consisting of 439 PzKw IVs, 623 PzKw 38Ts, 149 PzKw 35Ts and 1156 light PzKw Is and IIs.

The invasion commenced at 0315 on 22 June 1941. It was executed by Army Groups North, Centre and South, the first and last being spearheaded by one panzer group and Army Group Centre by two. The overall objective was a line running south from Archangel to Astrakhan on the Volga, with Moscow, the hub of the country's strategic rail network, as the key objective. Once again, the campaign is too well known to require more than a brief reminder of its course. The Red Army, whose reactions were far slower than had been those of the French, was repeatedly carved into pockets by the German spearheads and these were then forced to surrender to the follow-up infantry divisions. By the time the offensive ran down in December the Germans had taken 2,986,000 prisoners and captured 14,287 tanks and 25,212 guns; no accurate record exists of Soviet killed and wounded, nor of equipment destroyed in action. During the same period German losses amounted to 2300 tanks and approximately 800,000 personnel casualties. Yet, Barbarossa failed for a number of reasons. First, because of the campaigns in Greece and Yugoslavia, it had begun too late and was overtaken by the Russian winter. Then, Hitler failed to grasp that 100 miles' running in the Russian interior was very different to a similar journey over the excellent roads of western Europe, or that it made correspondingly higher demands on his tanks' machinery and his fuel supply. Finally, at the moment when the entire German effort should have been directed at securing Moscow, he intervened personally in the conduct of the campaign to divert resources against secondary objectives, thereby forfeiting priceless time that could never be recovered.

The new generation of Soviet tanks also provided a most unpleasant shock, although initially it was the KVs that made the greatest impression, as the following account, taken from the *German Report Series* of pamphlets, produced for the US Army in the post-war years by a team of German officers under the direction of Colonel General Franz Halder, graphically confirms:

'On 23 June 1941 the German Panzer Group 4, after a thrust from East Prussia, had reached the Dubysa and had formed several bridgeheads. The defeated enemy infantry units scattered into the extensive forests and high grain fields, where they constituted a threat to the German supply lines. As early as 25 June the Russians launched a surprise counter-attack on the southern bridgehead in the direction of Raseiniai with their hastily brought-up XIV Tank Corps. They overpowered the 6th Motorcycle Battalion which was committed in the bridgehead, took the bridge, and pressed on in the direction of the city. The German 114th Motorised Rifle Regiment, reinforced by two artillery battalions and 100 tanks (all belonging to the 6th Panzer Division), was immediately put into action and stopped the main body of the enemy forces. Then there suddenly appeared for the first time a battalion of heavy enemy tanks of previously unknown type.

The tanks overran the Riflemen and broke through into the artillery position. The projectiles of all defence weapons, except the 88mm Flak, bounced off the thick enemy armour. The 100 German tanks were unable to check the 20 enemy dreadnoughts and suffered loss. Several Czech-built tanks (PzKw 35Ts) which had bogged down in the grain fields because of mechanical trouble were flattened by the enemy monsters. The same fate befell a 150mm medium howitzer battery which kept on firing until the last minute. Despite the fact that it scored numerous direct hits from as close a range as 200 yards, its heavy shells were unable to put even a single tank out of action. The situation became critical. Only the 88mm Flak finally knocked out a few of the KV-1s and forced the others to withdraw into a wood.

'One of the KVs even managed to reach the supply route of the German task force located in the northern bridgehead and blocked it for several days. The first unsuspecting trucks to arrive with supplies were immediately set ablaze by the tank. There were practically no means of eliminating the monster. It was impossible to bypass it because of the swampy surrounding terrain. Neither supplies nor ammunition could be brought up. The severely wounded could not be removed to the hospital for the necessary operations, so they died. The attempt to put the tank out of action with the 50mm anti-tank gun battery, which had just been introduced at that time, at a range of 500 yards, ended with heavy losses to crews and equipment of the battery. The tank remained undamaged in spite of the fact that, as was later determined, it got fourteen direct hits. These merely produced blue spots on its armour. When a camouflaged 88 was brought up, the tank calmly permitted it to be put into position at a distance of 700 yards, and then smashed it and its crew before it was even ready to fire. The attempt of engineers to blow it up at night likewise proved abortive. To be sure, the engineers managed to get to the tank after midnight, and laid the prescribed demolition charge under the caterpillar tracks. The charge went off according to plan, but was insufficient for the oversized tracks. Pieces were broken off the tracks but the tank remained mobile and continued to block all supplies. At first it received supplies at night from scattered Russian groups and civilians, but the Germans later prevented this by blocking off the surrounding area. However, even this isolation did not induce it to give up its favourable position. It finally became the victim of a German ruse. Fifty tanks were ordered to feign an attack from three sides and to fire on it so as to draw all of its attention in those directions. Under the protection of this feint it was possible to set up and camouflage another 88mm Flak to the rear of the tank, so that this time it was actually able to fire. Of the twelve direct hits scored by this weapon, three pierced the tank and destroyed it.'

Throughout, the Russian infantry had remained passive spectators to the action. When German troops from the northern bridgehead began to threaten the rear of the Soviet corps the latter hurriedly abandoned the gains it had made during its counter-attack and pulled back across the river, losing many of its tanks in a swamp as it withdrew. Checks of this duration were rare and in most cases it was possible to deal with KVs by means of direct fire from the 88mms or heavy artillery, dive-bomber attacks or even placing demolition charges below the rear overhang of their turrets.

The advance of Field Marshal Gerd von Rundstedt's Army Group South was spearheaded by Panzer Group 1, consisting of the 9th, 11th, 13th, 14th and 16th Panzer Divisions, under the overall command of Colonel General Ewald von Kleist. Within days of the invasion starting, Kleist's group had become involved in what, until 1943, was the war's largest tank battle, a fierce if untidy struggle that raged for four days over a wide area near Brody.

Commanding the Soviet South-West Front was Colonel General Mikhail Kirponos, an officer of acute perception who was also possessed of immense energy and drive. He was one of the very few Soviet commanders who distinguished themselves during the war against Finland and he retained sufficient self-confidence to act on his own initiative. Under his command were several of new mechanised corps, either complete or in the process of forming, including the VIIIth with 600 tanks of which 170 were KVs and T34s; the IXth with 300 tanks, all of them older designs; the XVth with about 600 tanks including 133 KVs and T34s; the XIXth with only 160 tanks, of which only two were T34s; and the XXIInd, which was somewhat better off, although it lacked ammunition for its 31 KV-2s. Observing that Panzer Group 1 had begun to operate well ahead of the main body of Army Group South, Kirponos decided to concentrate his mechanised corps against it and isolate the Germans' armoured spearhead with converging attacks from the flanks. This, given the crippling disadvantages that the corps were not only located between 50 and 300 miles behind the front but were also themselves dispersed in their peacetime stations, coupled with the primitive state of Russian communications, was a task of Herculean proportions. Where no radio links existed, staff officers resorted to the civilian telephone system, if it was still working; if it was not, they delivered orders personally, using every means of transport available, including cars and elderly biplanes. Somehow, the whole cumbrous mass was set in motion, and it was then that the Russians' troubles really began.

The terrain caused them as many problems as it did the Germans. Forests and swamps were obvious obstacles, but even on superficially good, open going apparently stretching to the horizon the way ahead could suddenly be barred by a steep-sided ravine with a stream meandering sluggishly along the bottom. For these reasons, both sides established lanes of movement based on the existing roads, which were little better than cart tracks, snaking their way forward along them at five or six miles per hour. Having eliminated the Red Air Force during the early hours of the war, the Luftwaffe again enjoyed the freedom of the skies and its dive-bomber squadrons pounced eagerly on the long Russian columns to bomb and strafe. Frequent breakdowns, especially among the older designs, added to the trail of vehicles left behind, and since the recovery and spares system was as primitive as the radio communications network it was inevitable that, if the German advance continued, tank casualties would have to be abandoned and destroyed by their crews. The greater part of XV Mechanised Corps spent an entire day floundering in a bog.

In such circumstances operational timings became nothing more than a pious hope. Thus, although the Russian doctrine placed great emphasis on the importance of mass, when Kirponos' corps went into action they did so piecemeal with such of their units as had arrived first. Their lack of training and experience was immediately apparent. To quote the Halder team again:

'Tank attacks generally were not conducted at a fast enough pace; frequently they were not well enough adapted to the nature of the terrain, facts noted time and again throughout the entire war. The training of the individual tank driver was inadequate; the training period was apparently too short and losses in experienced drivers were too high. The Russian avoided driving his tank through hollows or along reverse slopes, preferring to choose a route along the crests which would give fewer driving difficulties. This practice remained unchanged even in the face of unusually high tank losses. Thus the Germans were in most cases able to bring the Russian tanks under fire at long range and to inflict losses even before the battle had begun. Slow and uncertain driving and numerous firing halts made the Russian tanks good targets. Premature firing on the Russian tanks, though wrong in principle, was always the German solution in these instances. If the German defence was ready and adequate the swarms of Russian tanks began to thin out very quickly in most cases.'

Throughout this series of largely uncoordinated engagements the Russians' lack of radios meant that each tank company tended to bunch around its commander, watching for his hand or flag signals. In these circumstances it was an easy matter for the German gunners to identify the command tank and eliminate it early in the engagement, leaving the Russians leaderless. The KVs and T34s presented an unwelcome problem - one T34 is recorded to have shrugged off twenty-four 37mm armour piercing rounds and only retired when the next round jammed the turret – but were increasingly isolated when the Germans concentrated on whittling away the older T26s, T28s and BTs, leaving them to be destroyed by the means described above.

The climax of Kirponos' effort began on 26 June. VIII and XV Mechanised Corps struck north from Brody while IX and XIX Mechanised Corps drove into Panzer Group I's northern flank, the intention being that the two thrusts should meet in the area of Dubno. During the evening of 29 June, Lieutenant General Siegfried Heinrici's 16th Panzer Division was checked frontally then assailed from both flanks.

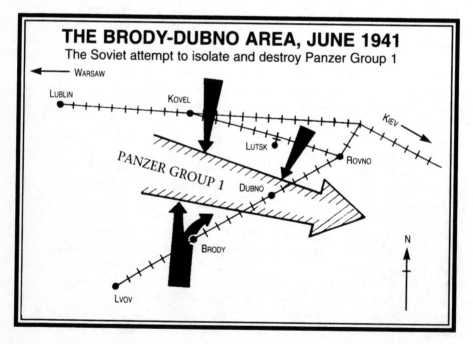

THE BRODY-DUBNO AREA, JUNE 1941
The Soviet attempt to isolate and destroy Panzer Group 1

'At 1800 the division abandoned the numerous small battles in which it had become involved and redeployed to attack Werba, south-east of Beresteczko,' wrote the historian of the 2nd Panzer Regiment. 'Ort was quickly taken but a few kilometres beyond the leading companies were halted by heavy tank and anti-tank fire from positions on the hills north of Plaszowa, where the enemy had dug in. Heavy fighting continued until dusk, with the leading platoons having to send back individual tanks to replenish their ammunition. Suddenly, alcohol-inflamed Russians emerged from the marshland on either side of the highway, yelling "Hurra!" and shooting wildly, while enemy tanks appeared on the road itself. The rest of the divisional battlegroup, personnel carriers, artillery and flak, closed up and completely unprotected on the road, now found itself in serious trouble. Vehicles, guns and damaged tanks struggled to turn and retire. Finally, the order to disengage was given and in something approaching panic the whole battlegroup headed for the rear. After only a short period, however, the fire of individual tanks and guns succeeded in halting the enemy advance.'

Darkness also prevented the Russians exploiting their success. As their tanks withdrew they were followed closely by part of 2nd Panzer Regiment's 1st Battalion, commanded by the remarkable Major Count Hyazinth von Strachwitz, of whom more anon. Undetected, the German tanks entered the Russian lines and, as soon as there was sufficient light to shoot, began to create mayhem among their opponents, destroying the batteries that had held up the previous day's advance. This and other similar actions was to earn Strachwitz the coveted Knight's Cross. Meanwhile, the rest of 16th Panzer Division, having been reinforced with infantry, was able to prevent the two converging prongs of Kirponos' counter-offensive from meeting.

By the end of the month the Battle of Brody-Dubno was over. Kirponos extracted the remnants of his shattered corps and withdrew towards Kiev, fuel shortages and breakdowns contributing to the trail of abandoned tanks that marked his route. The reinforced 34th Tank Division, which performed annually in Red Square on May Day and was considered to be an élite, had been surrounded and destroyed during the fighting, while the tank strength of other formations was reduced to skeletal levels. Between 27 June and 1 July the 16th Panzer Division alone claimed the destruction of 293 tanks. As Kleist's panzer group resumed its advance it came across the abandoned hulks of older Soviet designs that had failed to get into action, including multi-turreted T35s and SMKs that were fearsome to behold but had small combat potential; unimpressed, the Germans called them Kinderschrecke, or Bogeymen.

On the other hand, Kirponos' counter-attack had also caused the Germans loss and delay as well as such serious concern that this was reflected in Halder's diary. Had he survived, Kirponos would undoubtedly have exercised considerable influence in the continuing development of the Soviet armoured corps; as it was, he was ordered by Stalin to hold Kiev to the last man and was killed in September while leading an attempt to break out of the huge pocket that had been formed around the city.

Stalin and his General Staff, however, were capable of learning from their mistakes. It was clear that not only, given the limitations of the Russian command and communications apparatus, could the mechanised corps not be controlled adequately, but also that the entire art of commanding armoured formations would have to be have to be learned from the ground upwards. As the campaign progressed, therefore, the basic formation became the tank brigade, consisting of approximately 50 tanks, a motor rifle battalion and a reconnaissance battalion. By the spring of 1942 two or more such brigades, with supporting arms, could be formed into tank corps, and some four months later it became possible to form tank armies from two or three tank corps, plus appropriate army troops. Again, while the principle of a single operational plan was strictly adhered to, reliable officers at various levels of the chain of command were permitted to insert limited local variations at their discretion. This gave the Russians some degree of flexibility, although it never approached that of the German Army.

By December 1941 the Soviet tank strength had fallen to 2000 vehicles. There was, however, no cause for immediate alarm as the worst winter in living memory had put an end to mobile operations. The Russians were also satisfied with the performance of their new generation of tanks and had decided to concentrate on the production of the T60/70 light, the T34 medium and the KV heavy. The factories themselves had been moved beyond the Luftwaffe's reach to new locations east of the Urals, where they had begun turning out fighting vehicles in numbers that dwarfed the capacity of the German war industry.

For its part, the German armoured corps had won its battles but, for the first time and for reasons beyond its control, it had failed to win the campaign. It was unnerving to discover that the enemy, so despised by Dr Goebbels' propaganda apparatus, actually possessed better tanks; Kleist himself had no hesitation in describing the T34 as the best tank in the world. Suddenly, the German tank crews had found themselves underarmoured and outgunned. At a stroke, every tank in German service, with the exception of the PzKw IV, had become obsolete; and, since the German Army, now used to success in short wars, was equipped in breadth but not depth, no second generation of tanks existed.

The immediate crisis was solved in a number of ingenious ways. The wide turret ring of the PzKw IV made it possible for the short 75mm howitzer to be replaced with a 75mm L/43 high velocity gun, followed in 1943 by the improved 75mm L/48. Both these weapons were also fitted to the Assault Artillery's assault guns. Obsolete and captured tank chassis were fitted with a lightly armoured superstructure, armed with German 75mm or captured Russian 76.2mm anti-tank guns, and issued to tank destroyer units. In the longer term, nevertheless, the provision of new tanks was a matter that could not be avoided. Suggestions that the T34 should simply be copied were rejected on political, technical and operational grounds. In fact, the basic specification for the new medium PzKw V, armed with a 75mm L/70 gun, was issued as early as November 1941. This vehicle, better known as the Panther, was rushed through its development phase and entered service in the summer of 1943. In the meantime, work had been continuing on a pre-war project that actually lay outside the mainstream of the German armoured corps' operational concepts, namely a 54-ton breakthrough tank with an 88mm L/56 gun capable of destroying any tank in the world. Enraged by the stubborn resistance encountered at Leningrad, Hitler ordered that production of the vehicle should be accelerated with the result that the PzKw VI or Tiger E reached the front in August 1942 and was issued to independent heavy tank battalions. An even heavier design, the 69-ton Tiger B, sometimes known as the Royal or King Tiger, was armed with an 88mm L/71 gun and began leaving the production lines in February 1944.

The Russian response was first to increase the calibre length of their 76.2mm guns from L/30.5 to L/41.2, then upgun the KV with an 85mm L/43 weapon. The same weapon was chosen to upgun the T34, although the process required a completely redesigned three-man turret, and the result entered service in the spring of 1944. Meanwhile, the IS heavy tank series, armed with a 122mm L/43 gun, had been developed from the KV and was also leaving the factories.

As each side continued to fit more and more powerful guns, the other responded by protecting its tanks with thicker and more scientifically arranged armour. Of necessity, these developments were also absorbed by the British and US Armies. In 1939 medium and cruiser tanks had gone to war protected by frontal armour which averaged 30mm, while for heavy and infantry tanks 78mm was regarded as impressive. By 1945, however, the situation had altered dramatically. The Panther had 80mm frontal armour; the Tiger E 110mm; the Tiger B 185mm; the T34/85 75mm; the IS-II 160mm; the Comet 101mm; the Churchill VII 152mm; the Sherman 105mm; and the Pershing 110mm. The tanks of 1945 bore little resemblance to those of 1939 and, terrain permitting, engagements between them had been fought at progressively longer ranges for the previous three years.

Although the beginnings of this continuous process, sometimes referred to as the gun-armour spiral, were apparent before the German invasion of the USSR, it was the 1941 tank battles on the Russian Front that provided it with an unstoppable momentum.

Desert Death-Ride:
Tel el Aqqaqir, 2–4 November 1942

At 2140 on 23 October 1942 the night sky above Egypt's Western Desert was split by a gigantic series of flashes as 592 guns opened fire simultaneously, signalling the start of the Second Battle of Alamein. For more than two years the struggle had raged up and down the desert, twice reaching as far west as El Agheila in Libya, without either side, the British and Commonwealth divisions of the Eighth Army on the one hand and the Germans of the Deutsches Afrika Korps (DAK) and their Italian allies on the other, being able to inflict a decisive defeat on the other.

One frequently made but nevertheless apt analogy of the combatant armies during the North African campaign was that they resembled two boxers, each of whom was tied to his own corner by the elastic rope of his own supply lines, without which, of course, he would be unable to function at all. The further one boxer advanced from his corner, the stronger became the pull exerted by the rope and the fewer aggressive moves he was able to make, while for his opponent the reverse was true. Then, in May and June 1942, the British sustained a serious reverse at Gazala and were bundled back into Egypt. Worse was to follow, for in the immediate aftermath of the battle, Tobruk, which had been held with dogged determination throughout the previous year's campaigning, was quickly stormed and forced to surrender, earning General Erwin Rommel, the commander of Panzerarmee Afrika, his field marshal's baton.

By now the personality of Rommel had begun to take an unhealthy hold on his opponents' imaginations. In the mind of the average British and Commonwealth soldier, whatever his generals did, Rommel would eventually get the better of them. Rommel did indeed possess the ability to predict the probable speed of his enemies' reactions and, by exercising forward command he was able to gather together such forces as were immediately to hand and ruthlessly exploit any weakness he detected. In this he was assisted by the fact that he kept his own armoured divisions concentrated whereas the British armoured brigades often fought uncoordinated actions without the full support of other arms and were therefore defeated in detail. Yet, throughout his career Rommel had always taken risks that came dangerously close to gambling. Thus far he had been lucky, largely because he had been saved from the worst consequences of his actions by his opponents' mistakes. His opening moves during the Battle of Gazala, for example, were badly conceived and resulted in the DAK, severely mauled and shaken, being pinned with its back against the British minefields; only the British failure to exploit the situation promptly enabled him to regroup and regain the offensive, with devastating results.

In the euphoria following his victory at Gazala, the capture of Tobruk and his subsequent promotion, he made a strategic error that was to cost him the campaign. He decided that he would pursue the broken Eighth Army deep into Egypt, denying it the space and time in which to reorganise, with the ultimate object of breaking through its remnants to the Nile and the Suez Canal. If this was a product of self-generated hubris, it was also the sort of sweeping concept that appealed to Hitler, who promptly gave his approval. Other senior German officers, notably Field Marshal Kesselring, recognised that Rommel not only lacked the physical resources to complete his grand design, but also that, in accordance with the economic rules of desert warfare, the British would become progressively stronger the closer they withdrew to their bases.

Jumbled together, the two armies straggled eastwards to a village named El Alamein, lying on the railway some 60 miles west of Alexandria. At this point the width of negotiable desert was reduced to a mere 35 miles, bounded on the south by the shifting sands and salt marshes of the Qattara Depression and in the north by the sea. Here, under the personal command of General Sir Claude Auchinleck, the Commander in Chief Middle East, the Eighth Army turned and in the month-long series of engagements which have become known as the First Battle of Alamein, fought the Germans and Italians to a standstill.

Suddenly, Rommel found himself held fast within a strategic straitjacket. The RAF and the Royal Navy's Malta-based submarines were preying upon his seaborne communications with Italy to such good effect that most of his supplies were routed through Tripoli, 1000 miles to the west, with the result that such fuel as reached the front barely kept pace with daily requirements. The situation would have been even worse if it had not been for the Western Desert Railway, which had been extended to Tobruk. But even the railway, while it saved precious fuel that would otherwise have been consumed by motorised supply convoys, was a tenuous link, for it was subjected to regular air attacks and raids by the SAS and could not be used east of Mersa Matruh. Simultaneously, Rommel was aware that the Eighth Army was being reinforced at a rate he could not hope to match. Nevertheless, by the end of August he had accumulated just sufficient resources to attempt a renewal of his drive to the east, using the same right hook with which he had begun the Gazala battle. His attack was sharply repulsed at Alam Halfa ridge and he was forced to retire behind his own minefields, the net result of the operation being that he was held even tighter within the straitjacket. He was unable to take the offensive, yet to withdraw would involve a mobile battle which he lacked the fuel to sustain, as well as being politically unacceptable. The only remaining alternative was to hold his position behind thickening minebelts and await the Eighth Army's inevitable attack.

Rommel had noticed that at Alam Halfa the Eighth Army fought a more tightly controlled and coordinated battle than had been the case for many months. This reflected the temperament of its new commander, Lieutenant General Bernard Montgomery, who had arrived from England to find it tired, cynical and unsure of eventual victory. Montgomery immediately decided that his first task was to eradicate the image of Rommel from the minds of his men and replace it with his own. Substituting the black beret of the Royal Tank Regiment for his general officer's cap, he toured every fighting formation under his command and in his brusque manner told them that once the necessary preparations had been completed he intended kicking Rommel and his Axis army out of Africa once and for all. Something in his manner told the troops that he meant business and they responded. Next, like a cleansing wind, he swept away heretical practices such as dispersion, beloved of some old desert hands, which had exerted so baleful an influence on the Army's tactics for at least a year, and replaced them with a strict orthodoxy demanding concentration and sustained cooperation between arms at every level; those unwilling or unable to adapt were ruthlessly removed from their commands. The success of these methods at Alam Halfa, together with the intensive training programme that followed, restored the Eighth Army's self-confidence and belief that it would win.

Meanwhile, reinforcements continued to pour into Egypt. The news of the Army's defeat at Gazala and the fall of Tobruk reached Churchill while he was visiting Roosevelt in Washington and had left him profoundly shocked. With typical generosity the American President had offered to send the US 2nd Armored Division to Egypt. However, when General George C. Marshall, the US Army's Chief of Staff, pointed out that it would take up to five months for the division to reach the Middle East and become fully integrated it was decided to send 300 Shermans instead, plus 100 M7 (Priest) 105mm self-propelled howitzers. These were despatched on 15 July and arrived in Egypt shortly after Alam Halfa.

A persistent belief existed among the Eighth Army's tank crews that they were outgunned in the tank battle. This had never been altogether true, although on the eve of Second Alamein the Valentine infantry tanks and earlier models of the Crusader cruiser tank were still armed with the two-pounder, and the Stuart with the 37mm, both of which were outranged by German tanks armed with weapons of 50mm calibre and above. On the other hand, the 75mm guns of the Sherman and the Grant (an American M3 Lee with an upper turret built to British specifications, first committed to action at Gazala) outranged every German tank save the 75mm L/48-armed PzKw IVF2, of which Rommel possessed comparatively few. This belief had its roots in the heavy tank losses inflicted by the enemy's 'sword and shield' tactics, in which the Germans expertly drew their opponents within range of all-but invisible anti-tank gun screens. The most dangerous component of these was the high-standing 88mm dual-purpose anti-aircraft/anti-tank gun, which was capable of penetrating any Allied tank at long range and possessed a fearsome reputation. The standard German 75mm anti-tank gun, being much smaller and therefore capable of easier concealment, was also a formidable weapon, as was the captured Russian 76.2mm, rechambered to take German ammunition. The German 50mm anti-tank gun could penetrate the Crusader and Stuart, but not the frontal armour of the Valentine, Sherman and Grant save at close range, and the same was true of the Italian 47mm anti-tank gun. By this stage of the campaign the significance of anti-tank gun screens in the German way of fighting had, of course, been long appreciated, but whereas formerly British armoured formations had been forced to rely on their own supporting artillery to neutralise them, the ability of the Sherman and the Grant to fire a high explosive shell enabled them to tackle the problem quickly with direct gunfire, although the inability of the Grant to fight fully hull-down because of its sponson-mounted main armament presented difficulties which could not be easily overcome.

Quantitatively, the Eighth Army could field over 1000 tanks, including 252 Shermans and 170 Grants, plus 200 in immediate reserve and approximately 1000 more in workshops. Against this, the Axis could deploy only 520 tanks, including 30 PzKw IVF2s and 88 PzKw III Model Js, the latter armed with a 50mm L/60 gun, the balance consisting of older PzKw IIIs and IVs and 278 Italian M13s, plus a further 32 in workshops. In other areas including armoured cars, field and medium artillery and anti-tank guns the British also possessed a marked numerical superiority, although they had nothing to compare with Rommel's eighty-six 88mm guns. In the air the RAF, recently reinforced by a number of USAAF squadrons, outnumbered the Luftwaffe and the Regia Aeronautica in the ratio of five aircraft to three.

Nevertheless, the task confronting Montgomery was far from easy. The Axis army's protective minebelt was between two and five miles in depth and along the front German units were interspersed with Italian to provide a stiffening. To avoid the worst effects of the British bombardment the infantry had been ordered to hold the forward line with as few troops as possible and base their main position well to the rear. If the British armour did succeed in breaking through the minefields it was to be halted by dug-in anti-tank gun screens and promptly counter-attacked by one or both of two armoured groups, consisting of 15th Panzer Division and the Italian Littorio Armoured Division in the north and the 21st Panzer Division and the Italian Ariete Armoured Division in the south, before it could deploy its superior numbers. By 23 September there was nothing more that Rommel could do and, exhausted, he temporarily handed over command of Panzerarmee Afrika to General Georg Stumme and left for sick leave in Germany.

Montgomery, well aware of German anxieties regarding fuel, was determined to exploit these to the full by fighting what he described as a 'crumbling' battle, attacking first on one sector and then another, thereby forcing the Axis armour to react and burn up its priceless supplies in constant movement. His own army consisted of three corps. XXX Corps, commanded by Lieutenant General Sir Oliver Leese, was deployed on the northern sector of the line and consisted of the 9th Australian Division, 51st Highland Division, 2nd New Zealand Division containing two infantry brigades and 9th Armoured Brigade, 1st South African Division, 4th Indian Division and 23rd Armoured Brigade Group, a specialist infantry support formation with 194 Valentines. Immediately behind XXX Corps and overlaying its rear areas was Lieutenant General H. Lumsden's X Corps containing the 1st Armoured Division (Major General R. Briggs) and 10th Armoured Division (Major General A. H. Gatehouse). On the southern sector of the front was Lieutenant General B. G. Horrocks' XIII Corps, consisting of the 7th Armoured Division, the 50th and 44th Infantry Divisions, 1st and 2nd Fighting French Brigade Groups and the 1st Greek Infantry Brigade Group. Montgomery's plan required XIII Corps to mount holding attacks while XXX Corps delivered the main blow in the north, securing two corridors in the minefields through which X Corps would pass and write down the Axis armour.

In the event, despite the Eighth Army's high standard of training, this proved to be more difficult to achieve than many had anticipated, partly because of congestion in the minefields gaps, and partly because of the enemy's tenacious resistance. The course of Second Alamein was marked by several fiercely contested battles within the battle, notably the first major clash of armour in the area of the Meteiriya and Kidney Ridges, the defence of Outpost Snipe and the Australians' epic struggle around Thompson's Post on the coastal sector, but little ground was gained and the Axis army seemed as firmly entrenched as ever. By 29 October real concern was being expressed in the War Cabinet about the apparent lack of tangible results, but Montgomery had calculated almost exactly the time required to destroy his opponent and was not unduly worried. In fact, Panzerarmee Afrika had been hurt very badly and was slowly bleeding to death, just as he intended. Stumme had sustained a fatal heart attack on the 24th and Lieutenant General Ritter von Thoma, Commander of the DAK, had assumed command until Rommel returned to the front the following evening. To the alarm of his senior officers, Rommel persisted in committing his armour to counter-attacks that merely eroded its strength and consumed the vital fuel reserve, which never amounted to more than one or two days' running. With regard to the latter, he had been promised 4244 tons of fuel between 27 October and 1 November but such were the depredations on his supply line that only 893 tons actually arrived. By the evening of 28 October he had already begun to consider the possibilities of a tactical withdrawal to Fuka.

On the same day that members of the War Cabinet were voicing their doubts in London, Montgomery was planning what he intended to be the death stroke to Rom-

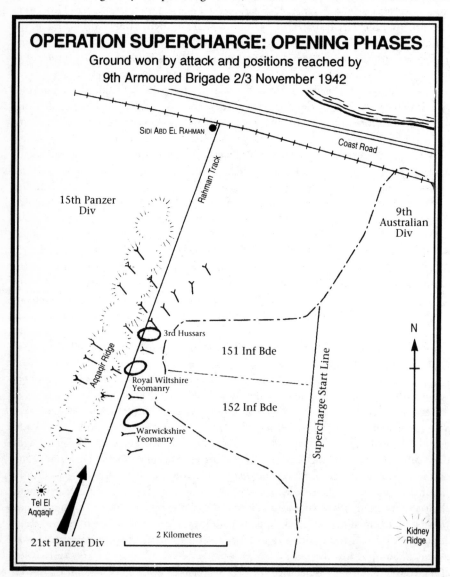

OPERATION SUPERCHARGE: OPENING PHASES
Ground won by attack and positions reached by
9th Armoured Brigade 2/3 November 1942

mel's army. Codenamed 'Supercharge', this would be delivered north of Kidney Ridge and consist of three phases:

> *Phase 1*, breaking into the enemy's forward defences on a two-brigade frontage east of the Rahman Track, his major north-south means of communication;
> *Phase 2*, destroying his anti-tank gun screen and breaking through his main position on Aqqaqir Ridge, west of the Rahman Track;
> *Phase3*, bringing to battle and destroying his armour, then breaking out into the open desert beyond.

The first phase would be executed by Major General Bernard Freyberg's 2nd New Zealand Division, with two regiments of 23rd Armoured Brigade under command. By this stage of the battle Freyberg's division was critically short of infantry and the assaulting infantry brigades were provided by 51st (Highland) Division and 50th (Northumbrian) Division; respectively, 152nd Brigade (Cameron and Seaforth Highlanders) supported by 50 RTR on the left, and 151st Brigade (Durham Light Infantry) supported by 8 RTR on the right.

The second and most difficult phase would by carried out by Brigadier John Currie's 9th Armoured Brigade (3rd Hussars, Royal Wiltshire Yeomanry and Warwickshire Yeomanry). The third phase was the responsibility of Briggs' 1st Armoured Division, containing two armoured brigades, Brigadier A. F. Fisher's 2nd (The Queen's Bays, 9th Lancers and 10th Hussars) and Brigadier E. C. N. Custance's 8th (3 RTR, Nottinghamshire Yeomanry, better known as the Sherwood Rangers, and Staffordshire Yeomanry).

As the operation would result in the climactic tank battle of the Desert War, to be named after Tel el Aqqaqir, the highest point on the enemy ridge, and was fought by representative elements of the entire Royal Armoured Corps, including Regular and Territorial units of the Royal Tank Regiment, Regular cavalry regiments and Yeomanry regiments, a word or two of clarification is necessary for those unfamiliar with British tactical practice and tradition.

Because of its specialist role 23rd Armoured Brigade, commanded by Brigadier G. W. Richards, fought with its units dispersed among the infantry divisions for which it provided close support. It consisted of one Regular regiment, 8 RTR, and three Territorial regiments, 40 and 46 RTR from Liverpool and 50 RTR from Bristol, raised from infantry battalions shortly before the outbreak of war. The brigade was equipped throughout with the Valentine infantry tank which, while undergunned, was reliable, stoutly armoured and small enough to be easily concealed. In order to keep casualties to a minimum, infantry attacks were made at night and during these the tanks maintained the closest possible contact. Their most important task, however, was to remain on the captured objective and beat off the enemy's counter-attacks, which were often delivered at about first light, until the infantry's own anti-tank guns could be brought forward through the minefield gaps and dug in. The success of these tactics generated a mutual confidence between infantry and tanks that had been lacking for many months, and their effect was duly noted and recorded by Rommel himself: 'The storming parties were accompanied by tanks which acted as mobile artillery, and forced their way into the trenches at the point of the bayonet; everything went methodically and according to a drill.'

The rest of the armoured brigades taking part in Supercharge fought as organic tactical formations. Each of their regiments contained one light squadron equipped with two-pounder and six-pounder Crusaders, and two heavy squadrons equipped with Grants and Shermans; with the exception of 2nd Armoured Brigade, the heavy squadrons of which consisted entirely of Shermans. No fewer than four of the nine armoured regiments were Yeomanry, a term with which non-British readers may not be familiar. Most Yeomanry regiments had been raised as volunteer cavalry units for home defence during the French Revolutionary and Napoleonic Wars, being recruited from gentlemen farmers and others who could provide their own mounts and accoutrements. Some had been called out to assist the civil power during the troubles of the Industrial Revolution, and all had sent contingents to South Africa during the Second Boer War. During World War I the majority had fought mounted in various theatres of war. Since then, many had been transferred to different arms of service but a handful had retained their horses and on the outbreak of World War II had served in Palestine with the 1st Cavalry Division before converting to armour. Being mostly countrymen, their common interest in field sports was often reflected in the metaphors they used in action and, as these activities also brought them together socially within their counties, many were already friends of long standing or related by marriage. Their regiments, therefore, possessed a closer sense of family than was usual even in the British Army and, because of this, their casualties hurt them deeply. Those about to be committed to Supercharge had fought at Meteiriya Ridge earlier in the battle, sustaining heavy losses, and these debts they were anxious to repay.

However, at the operational briefing held by Freyberg it was immediately apparent that the mission given to Currie's 9th Armoured Brigade, namely to overwhelm the enemy's anti-tank gun screen, amounted to nothing less than a death ride. Currie, a former horse artilleryman who gave his subordinates a hard time during training but mellowed somewhat in action, pointed out that losses amounting to 50 per cent could be expected. Freyberg's answer, delivered in a matter-of-fact tone, stunned the conference into total silence:

'It may well be more than that. The Army Commander has said that he is prepared to accept 100 per cent.'

The news was communicated to Currie's regiments, but was kept from those below the rank of squadron commander. Lieutenant Colonel Sir Peter Farquhar, commanding the 3rd Hussars, requested and was granted a personal interview with Montgomery. 'The Galloping Third' had been in the Desert War almost from the beginning, taking part in the defeat of the Italian Tenth Army at Beda Fomm and many other actions. In the process it had incurred such heavy losses that very few of the regiment's pre-war officers had survived; again, earlier that year the regiment had sent a squadron to the Dutch East Indies where it had fought to the last tank against the Japanese. It was with these matters and the question of the Third's continuing existence in mind that Farquhar told the Army Commander that the task was 'just suicide'. Montgomery, however, had calculated the price he was prepared to pay and replied that the job had got to be done, even if it cost every tank and crew. Later, Farquhar was to comment, 'I have always admired Montgomery for this frank reply – tough but typical of him. There was, of course, no more to be said'.

Supercharge was to have commenced during the night of 31 October but, for a variety of reasons, it was postponed until the early hours of 2 November. As the sun set the previous evening the troops involved moved into position and began the long, nerve-stretching period of waiting that precedes every operation, described so well by the historian of the Royal Wiltshire Yeomanry.

'Under such circumstances a couple of hours can seem like an eternity. The bravest of these men must have some qualms. It is bad enough for those who are more or less ignorant of what is in store for them, but for those who know the whole story it is a severe mental strain. Yet, on the whole, the impression of their deportment is one of unnatural calm. Most of them have climbed out of their tanks and are sitting in small groups chatting or leaning against them. Some betray their feelings by small nervous gestures – the drumming of fingers or the aimless rolling or unrolling of little bits of paper. Some chew sweets. More than one has a good swig at a flask. A few rather forced jokes are cracked and greeted with equally forced laughter. As the overanxious glance furtively at their wrist watches they see that the longer hands are catching up with the shorter on their journey to the appointed hour....Generally speaking, a man's thoughts at times like these are on the job in hand. The trooper is concentrating upon his immediate orders. The tank driver is probably engrossed in the behaviour of some little gadget which he fixed last night; he is praying that it will hold out until the battle is over. The officer has a number of worries of a different nature....In any case, whatever they are thinking, the time passes just the same.'

At 0105 the Durhams and the Highlanders, supported by their Valentines, began moving forward behind a rolling barrage while the Sappers toiled to complete minefield gaps. Both brigades sustained casualties but took large numbers of prisoners. By 0530 a hole had been punched in the enemy's forward defences and the break-in phase was complete. It was symptomatic of the Axis fuel shortage that a number of dug-in tanks were discovered in the captured positions. Unable to fight closed down at night, some were stalked by the infantry, who lobbed grenades into their turrets, or knocked out by the Valentines; a few, abandoned by their crews, were captured intact.

The second phase of the operation required 9th Armoured Brigade to pass through the newly established front line at 0545 and advance behind a rolling barrage to overrun the anti-tank gun screen covering the enemy's main position, beyond the Rahman Track; this would at least give the tanks the cover of the remaining darkness and place the enemy gunners at a disadvantage.

While the infantry had been fighting their way forward, Currie's regimental groups were moving along three parallel tracks that had been cleared through the minefields, the 3rd Hussars on the right, the Royal Wiltshire Yeomanry in the centre and the Warwickshire Yeomanry on the left. Each group, in addition to the armoured regiment itself, contained an infantry company from the 14th Sherwood Foresters, an anti-tank gun troop and a squadron from the New Zealand Divisional Cavalry Regiment, equipped with Stuarts. This approach march was itself extremely difficult, as at this point the desert had quite literally been pulverised by the passage of hundreds of tracked vehicles and the impact of thousands of shells. On the tracks the dust lay fully a foot deep, rising in a dense cloud which enveloped the vehicles and compounded the almost total darkness, penetrating eyes, ears, noses and mouths. Amid the constant stopping and starting tanks cannoned into each other and tempers frayed.

Currie came through on the radio to the Royal Wiltshire's commanding officer, Lieutenant Colonel Alistair Gibb, informing him that an enemy tank had been spotted lying just off one side of the track. 'What are you doing about it?' snarled the voice in the headsets. Gibb sent two of his own tanks to investigate, the commanders of which came perilously close to cutting Currie down with their machine-guns when he suddenly emerged from behind what transpired was a derelict. Brigadiers, reflected Gibb, should not be allowed in battle; they made life more dangerous for everyone else.

On the right the enemy had the range of the 3rd Hussars' track and had begun to shell it heavily and accurately. This resulted in all the wheeled vehicles being knocked out, serious casualties among the Foresters' company and the anti-tank troop, which lost all their officers, and the destruction of the engineers' masked lights indicating the route forward. As a result of this, the Hussars lost ten tanks which either strayed off the track into the minefields on either side or were damaged by shellfire. Nevertheless, they reached their Phase 2 start line on time, as did the Royal Wiltshires.

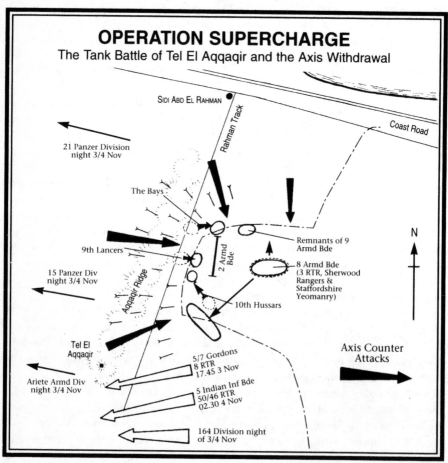

OPERATION SUPERCHARGE
The Tank Battle of Tel El Aqqaqir and the Axis Withdrawal

Sidi Abd El Rahman

Rahman Track

Coast Road

21 Panzer Division
night 3/4 Nov

The Bays

9th Lancers

15 Panzer Div
night 3/4 Nov

Aqqaqir Ridge

2 Armd Bde

Remnants of 9 Armd Bde

8 Armd Bde
(3 RTR, Sherwood Rangers & Staffordshire Yeomanry)

10th Hussars

N

Tel El Aqqaqir

Ariete Armd Div
night 3/4 Nov

5/7 Gordons
8 RTR
17.45 3 Nov

5 Indian Inf Bde
50/46 RTR
02.30 4 Nov

164 Division night of 3/4 Nov

Axis Counter Attacks

Unfortunately, the Warwickshire Yeomanry, commanded by Lieutenant Colonel Guy Jackson, did not. The regimental group had been directed into a false lead in the minefields and in sorting out the tangle four of its tanks had been immobilised with mine damage. By the time it reached the Phase 2 start line at 0545 Currie had altered the brigade plan. Farquhar and Gibb had urged him to let them proceed with their own regiments, thereby taking advantage of what little remained of the darkness, but Currie's concern was to smash as large a hole as possible in the enemy anti-tank gun screen using the entire brigade. For good or ill, therefore, the attack was postponed until 0615 and the timings of the barrage were adjusted accordingly.

The three regiments shook out into their squadron assault formations and as the shells began to scream overhead they moved forward in line abreast, the mechanical heirs to a score of cavalry charges, led personally by a brigade commander who would spare no one, least of all himself. It was soon apparent that the enemy's forward defence zone had been far deeper than had been anticipated, for in front of the tanks appeared the running shapes of German infantry, bolting to the rear or out of the path of the churning tracks. Some fell victim to the chattering machine-guns, some were run down, and others surrendered and were thumbed to the rear; of the last, a number returned to their weapon pits after the tanks had passed on and, until they were finally eliminated, were to keep the Foresters fully occupied . The tanks themselves kept as close to the rolling barrage as possible, some actually driving through it on the principle that the risk was justified by the prospect of being able to overwhelm the anti-tank gunners while they were still pinned down.

The 3rd Hussars, led by Major Michael Eveleigh's B Squadron, reached the Rahman Track and swept into the enemy, just as their forebears had swept into the ranks of the Sikhs at Chilianwala. Two four-gun anti-tank troops, armed respectively with German 50mm and Italian 47mm guns, were destroyed with gunfire at close range and the trails of their weapons were crushed beneath the tracks. Eveleigh was directing his troop leaders to attack a third battery when, with all the dramatic suddenness of the desert, the eastern sky began to lighten, changing colour rapidly from deep purple to red and from red to yellow. The British tanks were immediately silhouetted starkly against the dawn and laced with a hail of slashing shot from the German 75s and 88s, deployed deeper within the enemy position on the shallow slopes of Aqqaqir ridge. Eveleigh's own tank was hit and set ablaze at point blank range by a troop of Italian 47mm guns. He bailed out with his turret crew and, until the armour plate became too hot to touch, desperately struggled to free the jammed hatches of his driver and co-driver, to no avail. Aware that the Italian gunners were shooting at his gunner and operator with small arms, he emptied his revolver at them. At this point Lieutenant Charles Dorman, one of his troop leaders, seeing what was happening, attacked the Italians from a flank and wiped them out. The rest of the regiment had now come up and become heavily engaged in a series of personal close-quarter duels with the numerous gun positions and the leading tanks of 15th Panzer Division, which had just appeared to the north. As radio communications had broken down completely – the brigade had drawn replacement tanks just before the battle – Farquhar and Eveleigh had to walk round giving orders individually to each tank commander. During this phase of the action the regiment accounted for fifteen anti-tank guns, four field guns and five tanks, but by 0710 it had itself been reduced to seven tanks while only four of its officers remained alive and unwounded.

In the centre, the experience of the Royal Wiltshire Yeomanry was similar but worse. B Squadron's Crusaders, commanded by Major Tim Gibbs, actually crossed the Rahman Track and got among the guns before the enemy knew what was happening. For a few minutes the Yeomen did very much as they wanted, shooting down the gun crews and crushing the guns in their pits. Then, as the light strengthened, the squadron became the focus of concentrated fire and was quickly reduced to Gibbs' tank and that of Lieutenant Balding. Gibbs, detecting some distant movement to his left, thought at first that it must be the Warwickshires but as the tanks approached he suddenly realised that they were German. They belonged to a battlegroup of 21st Panzer Division and opened fire at 400 yards, penetrating the Crusader and killing its radio operator. Gibbs was wounded twice as he bailed out and, with his driver and gunner, made for Balding's tank, which was picking up survivors. Somehow, the Crusader survived as it retired under intense fire, although all of those who had scrambled aboard were killed or wounded, Gibbs being shot through the arm as he struggled to maintain his hold.

Meanwhile, the regiment's heavy squadrons, commanded by Lord Cadogan and Major Henry Awdrey, had come into action against the anti-tank guns, the muzzle flashes of which could be seen in a great arc around the brigade's frontage. As this part of the desert was completely flat it was impossible for the Grants and Shermans to take advantage of any sort of cover. The German armour also entered the battle, as did the enemy's heavy artillery, which began to hurl large calibre shell into the British ranks with increasing accuracy. Currie, standing erect in his turret, apparently impervious to the blizzard of flying metal around him, moved among them, tongue-lashing any commander who betrayed the slightest sign of nervousness. When Colonel Gibb, commanding the Royal Wiltshires, expressed concern about the havoc that was being wrought among his heavy squadrons, Currie told him that as 2nd Armoured Brigade was about to enter the battle he must hang on at all costs. Wondering what to do for the best, Gibb ordered both squadrons to put down smoke. For ten minutes the white screen brought a blessed relief, then the supply of smoke shells ran out and the carnage began anew. His own tank was among the first casualties and, pinned down in the sand by machine-gun fire beside the blazing, exploding wreck, he was forced to watch the enemy settling to the methodical destruction of his regiment. 'Dealing as always with one thing at a time he first of all takes on C Squadron. One by one he knocks them out. Then he liquidates what is left of Regimental Headquarters. Next, systematically, he turns to A Squadron. The survivors, pinned out in the open, fight back for all they are worth. But it can only be a matter of time.' In its self-sacrificial struggle the regiment inflicted losses that included fourteen anti-tank guns.

On the brigade's left the Warwickshire Yeomanry had taken a line of advance somewhat south of west, mistaking a rise south of Tel el Aqqaqir for the Tel itself in the darkness. A Squadron overran four Italian guns before the familiar pattern repeated itself and each tank found itself fighting its own individual battle against anti-tank guns, tanks and artillery. 'We kept on the move and belted away at the dug-in 75s and 88s,' recalled the driver of a C Squadron Grant. 'Tanks were brewing up all round us but we didn't get hit. There were flashes from guns on all sides and it wasn't until the sun came up that we knew which direction we were facing. We picked up two survivors from a 1st Troop tank and dropped them off by a gunpit near another tank. Then the Colonel came on the air: "For God's sake get those bloody guns before they get the lot of us!"'

The regimental radio net had been penetrated by an enemy intercept operator who evidently had a good view of the fighting as well as an excellent command of English, interjecting such unpleasant comments as 'Isn't it nice seeing all your beautiful new toys going up in smoke!' Colonel Jackson, whose tank, command pennants flying, continued to blaze away until it too was hit, could see that his men were also damaging the enemy and encouraged them in the language of the hunting field: 'The fox will break cover any moment now and the hunt will be on! Don't go back a yard!' When, some time later, Currie pulled back the sad remnants of his regiments a little way, the Warwickshires, who had entered the battle with 38 tanks, had been reduced to fourteen. Jackson, seriously ill with jaundice, collapsed shortly after; a man of immense physical and moral courage, he was to lose both legs during the Italian campaign but continued to ride to hounds after the war.

At about 0700 the regiments of Fisher's 2nd Armoured Brigade began to debouch from the minefield tracks, the Queen's Bays on the right, the 9th Lancers in the centre and the 10th Hussars on the left, to be confronted by a scene of utter desolation. To their front the 9th Armoured Brigade was burning, although here and there small groups of tanks were still firing defiantly. In the distance enemy tanks were also burning and smashed anti-tank guns, their muzzles poking skywards at impossible angles, littered the shallow slopes of Aqqaqir Ridge. Nearer to hand were smaller, equally vicious battles in which those German infantry who had been overrun but refused to surrender fought it out with the Foresters and dismounted tank crews.

In this place of sudden death Lieutenant Colonel Gerald Grosvenor, commanding the 9th Lancers, came upon Currie sitting on top of his turret. Grosvenor, who would become Duke of Westminster and was a nephew of the Duke whose armoured cars had operated on this very ground during World War I, climbed up beside him. Currie, bitter and angry, had expected 2nd Armoured to come through hard on his heels, and he remarked to Grosvenor: 'Well, we've made a gap in the enemy anti-tank screen and your brigade has to pass through, and pass through bloody quick!'

Grosvenor, surveying the battle raging to his front, thought he had never seen anything that looked less like a gap and said so. Currie's response was sharp and Grosvenor, feeling that this was neither the time nor the place for a row, returned to his own tank and reported the situation to Fisher.

The three regiments moved forward among the blazing derelicts of Currie's brigade and immediately became involved in the murderous gunnery duel. At 0740, however, the entire character of this changed dramatically. The sun emerged from the desert to the east 'like an enormous gunflash' and its piercing horizontal rays shone straight into the eyes of the German and Italian gunners, preventing them from taking effective aim. Now it was the British who began to score heavily.

Rommel, fully aware of danger presented by the salient that had been punched in his line, reacted as Montgomery hoped he would. For the rest of the day his four armoured divisions converged on the salient and mounted counter-attacks against its western, northern and southern edges supported by concentrated artillery fire. Simultaneously, the Luftwaffe's attempts to intervene in the battle were thwarted by Allied fighters who forced the Stuka dive-bombers to jettison their bombs over their own lines. Conversely, continued British attempts to break through were halted by stubborn resistance, despite the fact that the defenders of Aqqaqir Ridge were exposed to the fire of 360 field and medium guns and battered at regular intervals by Mitchell bombers.

Few of those involved in the fighting were aware of what was taking place beyond their immediate concerns. On the right of Fisher's brigade, the Bays, commanded by Lieutenant Colonel Alex Barclay, were engaged by 88s at 1000 yards as they moved forward to pass through the remnants of 3rd Hussars, losing two Shermans and a Crusader before an early morning mist caused both sides to cease firing. When this cleared the fire of both Sherman squadrons was directed at the guns opposite, around which the enemy crews could be seen reloading after each shot, and within 45 minutes had silenced them. At about 1000 heavy shelling erupted around the squadrons, preceding a counter-attack by some thirty tanks, the first of a succession that were to continue throughout the day.

'The enemy knew everything was at stake and did his utmost to drive back our armour, but our tank gunners shot magnificently,' wrote the regimental historian. 'Each time the German tanks came on one or two of them were hit, so that the others hesitated and stopped and none of their attacks could be pressed home. During the day's fighting, one of the hardest the Regiment ever had, almost every tank used up its complete load of ammunition.'

The historian of the 9th Lancers, who were pushing through what remained of the Royal Wiltshires, has left a vivid record of the action, in which he too commented that the day's fighting 'was about the worst we ever had.'

'All day we were fired at continually from three sides by 88s and 105s (sic). For hours the whack of armour piercing shot on armour plate was unceasing. Then the enemy tank attacks started. Out of the haze in serried lines they came, the low, black tanks. B and C Squadrons repulsed no fewer than six determined attacks and the regiment finished the day with a score of 31, of which 21 were set on fire. In addition, five guns were knocked out by putting air-bursts just over their pits. Passing the gunpits later we saw whole crews lying dead across their guns.

'The terrific cross-fire was taking its toll. Two of C Squadron's Shermans were burnt out, one cruiser destroyed and five other Shermans knocked out. Our own infantry, dug-in on the battlefield, suffered terribly, being killed by shells meant for us. In that torrent of shot and shell any man who moved was killed. We did what we could for them, but our attention was taken up in fighting for dear life.

'Time and again the tanks ran out of ammunition and as we could not afford to have one single tank out of the line the (supply) lorries had to rush forward to them across the shell-swept ground, taking what cover they could behind each tank.'

These lorry drivers, in constant danger of being blown to atoms if their unprotected vehicles were hit, were the unsung heroes of the battle, even if there were times when they must have wondered whether their hair-raising journeys had been strictly necessary. Corporal Cook of C Squadron reached Sergeant Harris's tank to find smoke billowing from the cupola. It was pleasantly aromatic and Harris emerged, puffing unconcernedly on his pipe. His field of fire was seriously restricted by the close proxim-

War on Wheels

Right: A section of Locker Lampson's Royal Naval armoured car unit in Galicia, 1917. The bulk of the unit's strength consisted of Lanchesters, although it did possess one Rolls Royce, seen in the foreground. The section's transport consists of one Ford and two Lanchester tenders, one of which is an ambulance. (Author's collection)

The Land Ironclads

Right: C Battalion Mark I male emerging from Chimpanzee Valley, late 1916. The timber and chicken wire construction on the roof was a defence against grenades. Trail wheels, of little use and vulnerable to shellfire, were soon dispensed with. The driver's hatch and the sponson doors have been opened to provide a current of air through the overheated interior of the vehicle. (RAC Tank Museum)

Blood on the Tracks

Below: Artist's impression of the Tank Corps about to break the German line at Cambrai. The picture correctly conveys the impression of concentration but the huge fascines have been omitted from the tanks' roofs. (RAC Tank Museum)

Left: The French Renault FT-17 was the first tank to be fitted with a fully rotating turret. (RAC Tank Museum)

Below: Captain Thomas Price's Whippet company scattering two German storm troop battalions at Villers-Bretonneux. (RAC Tank Museum)

Bottom: German view of an Allied tank attack south of Cambrai on 8 October 1918. The tanks are manned by the US 301st Tank Battalion, which lost ten of its twenty tanks during the action. (Imperial War Museum)

Above: The first tank versus tank battle: Villers Bretonneux 24 April 1918. A male Mk IV (Lieutenant Frank Mitchell, 1st Battalion The Tank Corps) with men of the 1st Battalion the Worcestershire Regiment in the trench. (By David Rowlands)

The Gembloux Gap

Right: The Char B, the principal equipment of the French Divisions Cuirassées, was heavily armoured as well as being armed with a 47mm gun in the turret and a 75mm howitzer in the bow plate. Although it was a formidable opponent and was known to the Germans as the Kolosse, its tracks and radiator louvres were both vulnerable; again, many went into action at the limit of their fuel endurance. (RAC Tank Museum)

Right: PzKw I (left) and PzKw IV (centre) of 1st Panzer Division, halted during the 1940 campaign in France. (Charles Kliment)

Barbarossa: Raseiniai and Brody-Dubno

Above: PzKw II and PzKw 38(T) of 8th Panzer Division in open leaguer, Russia, 1941. (Charles Kliment)

Below: In 1941 the combination of 76.2mm gun, well-angled armour and high-speed Christie suspension put the Soviet T34 a full generation ahead of German tank design. (Novosti)

Desert Death-Ride

Top: The M3 Stuart was the first American tank to serve in North Africa. (RAC Tank Museum)

Above: Grant squadron on the move in the Western Desert. The Grant was armed with a 37mm gun in the top turret and a 75mm gun mounted in a sponson. (RAC Tank Museum)

Below: The dreaded German 88mm dual-purpose anti-aircraft/anti-tank gun. (RAC Tank Museum)

Left: PzKw III in North Africa. Spare roadwheels and tracklinks have been attached to the front for additional protection. (RAC Tank Museum)

War to the Death

Left: The largest of the first generation of German tank destroyers was the Hornet, consisting of an 88mm gun mounted on a PzKw IV chassis. (Bundesarchiv)

Below: Panzer division on the march in Russia. At best only 10% of Panzer-grenadier units possessed the SdKfz 251 armoured personnel carriers shown here; the rest had to make do with unarmoured half-tracks and lorries. (Bundesarchiv)

Above: The PzKw V Panther became the best medium tank of World War II, but was committed at Kursk before its teething troubles had been solved. (RAC Tank Museum)

Below: The Tiger-equipped heavy tank battalions were used as 'fire brigades' to restore German fortunes whenever a breakthrough was threatened. This Tiger E was evidently abandoned near Florence; the Sherman on the left has lost a track to mine damage as it attempted to edge past. (New Zealand Official)

Top: Though crudely finished, the Soviet IS-II, armed with a 122mm gun, achieved parity with the Tiger E. (RAC Tank Museum)

Above: Towards the end of the war the assault gun and the Panzerfaust played an increasing part in the defence of Germany. (Bundesarchiv)

Hobo's Funnies

Left: Major General Sir Percy Hobart, commander of 79th Armoured Division. (Imperial War Museum)

Right: AVRE (Armoured Vehicle Royal Engineers) with fascine aboard; the launching sledge is visible beneath the fascine. The turret armour has been supplemented by spare track links. (Imperial War Museum)

Right: A Sherman Crab flailing its way through a minefield. (RAC Tank Museum)

Below: Buffaloes played a critical role in clearing the Scheldt estuary and the Rhine crossing. (RAC Tank Museum)

'Hideous Nights Pressed on Days of Horror'

Top left: A troop of M10 tank destroyers crosses a causeway of Churchill Arks beside a drowned Panther. (Imperial War Museum)

Centre left: South African Shermans firing in the role of supplementary artillery during the Italian Campaign. (South African National Museum of Military History)

Bottom left: A Churchill of the North Irish Horse in typical Gothic Line country. (Imperial War Museum)

The Armoured Car in World War II

Above right: German armoured car commanders plot their route forward. A photograph taken during the first year of the war, when the black beret was worn by all German armoured troops. (Imperial War Museum)

Below: 11th Hussar Morris towing home a captured Italian L3 tankette; North Africa, 1940. (Imperial War Museum)

Honour in Adversity

Left: The M18 tank destroyer, fast and well armed, was built to the Tank Destroyer Force's own specifications. (Imperial War Museum)

The Japanese Dimension

Left: Suicidal attacks on tanks by Japanese armed with explosive charges were a frequent occurrence. The threat could be reduced by fitting a stout wire grille over the engine deck, as shown here, and by providing a close escort of infantry. This Australian Matilda is shown clearing a jungle track to Slater's Knoll in Bougainville. (Australian War Memorial)

Left: An incident during the fighting on Bougainville. A Sherman and a rifle squad move forward to eject Japanese from positions they have infiltrated the previous night. (US Army Military History Institute)

Above and Beyond

Above: Australian walking wounded and prisoners take a breather beside the burned-out *Musical Box*, discovered some eight miles behind the original German front line. The full story did not emerge until after the war had ended. (Imperial War Museum)

Below left: Lieutenant Clement Arnold, the first officer to carry out an exploitation in a tracked fighting vehicle. A photograph taken in Freiburg prisoner of war camp shortly before the end of World War I. (Mrs Agnes Arnold)

Below right: Count Hyazinth Strachwitz von Gross Zauche und Caminetz, a member of the Upper Silesian aristocracy who was master of the tactical battle on the Russian Front. (Courtesy, Panzerbataillon 84)

Left: Up-gunned Israeli Sherman. (Camera Press, photo David Newell-Smith)

Left: Israeli Centurions in action. (Camera Press, photo Werner Braun)

Left: Knocked-out Egyptian T34/85 in Sinai. (Associated Press)

Right: A knocked-out T55 burns in the Sinai desert.

The Long War

Right: French paratroops provide a close escort for a Sherman as it moves through ideal ambush country in Vietnam. (SIRPA/ECPA)

Right: The ACAV (Armoured Cavalry Vehicle) was based on the M113 APC and possessed greater firepower than the light tanks of 1939. (USAMHI)

Green Screen Warfare

Above: A British Challenger main battle tank in Kuwait. Note the slabs of reactive armour, intended to defeat shaped-charge ammunition, bolted to the bow and side plates. (Lieutenant Colonel A. R. D. Shirreff)

Below: Iraqi T55 destroyed by B Squadron 14th/20th Hussars as it attempted to reverse out of its scrape. (Lieutenant Colonel A. R. D. Shirreff)

ity of the Bays and he did not require any ammunition. On the other hand, he was obviously pleased to see Cook, whom he genially greeted with: 'I say, you didn't bring any good books, did you?'

On the left the 'Shiny Tenth,' commanded by Lieutenant Colonel Jack Archer-Shee, had reached the minefield exit to find that the enemy had marked its position with smoke shells and was firing along fixed lines into it. The Regiment's B Squadron passed through the cloud and disappeared. Archer-Shee, unable to contact them on the radio, followed with the intention of delivering his orders personally. He found that the 9th Lancers had come up on the right and, spotting a distant group of enemy tanks in the half-light, mistook them for B Squadron. They opened fire at 500 yards, setting his own tank ablaze, fortunately without casualties to the crew, but were dispersed with the loss of four of their number when the rest of the regiment debouched from the minefield.

'Shortly after this an enemy tank counter-attack appeared over the crest of Tel el Aqqaqir. A and C Squadrons held their fire until the enemy tanks had reached a forward slope and then reaped a fine harvest which effectually stopped it and caused the enemy to retire to hull-down positions.'

By now the remnants of 9th Armoured Brigade, amounting to little more than a squadron, had come under the command of 1st Armoured Division and moved to assist in the defence of the northern flank of the salient, against which a heavy counter-attack was developing. 8 RTR's Valentines, which had remained with the Durham Light Infantry in the positions captured by the latter during the night, also went into action against the attackers.

Custance's 8th Armoured Brigade began reaching the increasingly congested battlefield at about 1000. The first regiment to emerge from the minefields was the Staffordshire Yeomanry, commanded by Lieutenant Colonel James Eadie, who observed that 26 German tanks had actually penetrated the northern edge of the salient, on the Durhams' flank. He immediately despatched his Crusader squadron to deal with the threat, followed by the heavy squadrons as they arrived. In a protracted exchange of gunfire the regiment destroyed three of the enemy's tanks, turned back his potentially dangerous probe and destroyed a number of anti-tank guns.

As they came up 3 RTR, under Lieutenant Colonel H. E. Pyman, and the Sherwood Rangers, under Lieutenant Colonel E. O. Kellett, a Member of Parliament, also deployed facing north and this may well have deterred further enemy efforts in this area. When, at 1037 Montgomery called HQ 1st Armoured Division, he seemed sufficiently satisfied with the situation to order Briggs, the divisional commander, to move 8th Armoured Brigade to the south-western sector of the salient. Custance despatched 3 RTR and the Sherwood Rangers to the position occupied by the 10th Hussars, who then closed up to the 9th Lancers, but as the Staffordshires were still heavily engaged he left them where they were until the afternoon. By comparison with events elsewhere, there was little activity in the brigade area, although the tanks were shelled constantly. Eadie's personal diary records that his Grant, *Defiance*, was twice struck by heavy calibre shells, one of which, exploding against the rear of the turret, produced a severe headache, the entry for the day concluding with a reflective glance in the direction of home: 'Should have been at Sudbury today – Opening Meet of the Meynell hounds.'

The sun set leaving both sides approximately where they had been at dawn. Superficially, the immense destruction of equipment and heavy loss of life had been for nothing. Excluding its casualties in the minefield, 9th Armoured Brigade had launched its attack with 94 tanks of which 75 had been knocked out, while of the 400 crewmen who manned them 230 had been killed or wounded. For all his iron image, Currie remained deeply disturbed by the losses among his men, a prey to thoughts that by postponing the attack he had contributed to these. Yet, had he gone in on time with two regiments and allowed the third to follow, he would have had to stand off the enemy's counter-attack alone for an additional thirty minutes; on balance, there was little to choose between the two alternatives. The success of the attack can be gauged by the fact that during day's heavy fighting 1st Armoured Division's losses amounted to only fourteen tanks written off and a further forty damaged.

In fact, unknown to most, the first crack had already appeared in the Axis front. Taking advantage of the morning mist and the shock caused by 9th Armoured Brigade's suicidal charge, the Royal Dragoons, an armoured car regiment, had worked two squadrons through the lines south of Tel el Aqqaqir. They had relied heavily on bluff, exchanging cheery waves with the uncertain German gunners, and were soon deep inside the enemy's rear areas without a shot having been fired at them. Here they spent the day shooting up transport and reporting movements before going into leaguer near El Daba.

General von Thoma's report to Rommel at 2015 that evening contained little but bad news. Seventy-seven German and 40 Italian tanks had been destroyed in the fighting. For the following day he could muster, at best, 35 German tanks; the Italians, whose thin-skinned M13s were unsuited to the sort of slugging match that had just taken place, had shown clear signs of being unsettled and could not be relied on. The strength of the anti-tank gun screen had been reduced by two-thirds. There were no more German reserves to commit to the battle. On the other hand, the line was holding, but only just. Rommel decided that the moment had come for him to start withdrawing such elements of his army as still possessed motorised transport, although this meant abandoning the rest. Their movement would be covered by the defenders of Aqqaqir Ridge.

On 3 November, therefore, Briggs and von Thoma continued their battle of mutual destruction, much as they had done the previous day, the experience of the Sherwood Rangers being typical.

'This morning, A Squadron were at the top of their form. They carried out a magnificent fighting reconnaissance, pinpointing enemy machine-guns and snipers' nests and then destroying them, and hammering away at the German anti-tank screen. Captain Bill McGowan hit a German 50mm anti-tank position with such niceness that the gun was blown clean up into the air. The squadron engaged the enemy field guns with indirect fire and spotted for our own 25-pounders who were also firing at them. When the squadron made things hot for an Italian battery of five field guns, the whole battery surrendered en masse. We were nobly helped by the infantry who, as we brassed up targets, nipped in from time to time and came back with parties of prisoners. Colonel Kellett reported to us: "The enemy are fighting a delaying action." The enemy line began to give way slowly and we moved out onto the Sidi Abd el Rahman track. To the north we could clearly see the Sidi (tomb of a saint) gleaming with its white walls and minaret. Between us and the Sidi were groups of German infantry sharply visible in their field-grey uniforms.

We had been making the pace a bit too hot for the Germans' plans, and now they ordered their anti-tank and field gunners to do their utmost to slow us down. The enemy fire grew fierce. Their redoubtable Eighty-Eights plastered us with high explosive and armour piercing shells. Captain Jack Tyrrell, our observation officer from the 1st Royal Horse Artillery, then organised a stonk. Our gunners certainly produced some accurate firing that day. We could see as some of the Eighty-Eights were put out of action by direct hits. Others were damaged and we saw vehicles drive up from the rear and tow them away. But the rest stayed on and shot on – those German gunners certainly had guts.'

On the Rangers' left the Staffordshire Yeomanry, fighting a similar action, had the unusual experience of seeing one of their tanks, hit and abandoned by its crew, returning home without human assistance.

Fighting under the command of the 10th Hussars on 3 November was an experimental unit known as Kingforce. Originally six tanks strong, this was equipped with the Churchill infantry tank and had first gone into action with the Bays at Kidney Ridge on 27 October, losing one of its number. At Aqqaqir it accounted for one tank and three anti-tank guns. Some measure of the intensity of the opposition can be gauged by the fact that one Churchill alone sustained thirty-one 50mm hits, offside track and idler damage, a dented six-pounder barrel and a smashed machine-gun mounting.

By the end of the day the British had gained ground but had still not achieved a breakthrough. Montgomery, aware that the enemy were thinning out and had begun to withdraw, decided that the area south of the Tel would be probed by infantry and Valentines. The first probe was to be made by 51st Highland Division and 8 RTR, directed at a point on the Rahman Track two miles south of the Tel. Shortly before the troops crossed the start line at 1745, Lieutenant General Sir Oliver Leese, commanding XXX Corps, telephoned the divisional commander, Major General Douglas Wimberley, and told him that 8th Armoured Brigade had actually reached the objective and that because of this there would be no supporting artillery barrage. Wimberley knew that this could not be the case and said so, but all the Corps Commander would concede was a scattering of smoke shells to mark the objective. The infantry, 5/7th Gordon Highlanders, therefore advanced riding on the Valentines, contrary to the normal drill used in contact with the enemy, with disastrous results. They ran into a storm of anti-tank and machine-gun fire which struck the exposed Highlanders, knocked out nine Valentines and damaged eleven more, leaving only twelve fit for action. The Gordons, having sustained 90 casualties, dug in short of the objective.

The second probe was also aimed at the Rahman Track, two miles further south. It was made by 5th Indian Brigade, with the support of 50 RTR, and commenced at 0230 on 4 November in the wake of a heavy barrage. The objective was secured with very little fighting and exploitation for 4000 yards beyond revealed that the enemy had gone. Through the gap passed the remaining squadron of the Royals and the 4th South African Armoured Car Regiment, to be followed by 7th Armoured Division, which had been ordered north from the southern sector of the front. At first light the Tel itself fell to an infantry attack by 7th Argyll and Sutherland Highlanders.

Sunrise also confirmed that the enemy had abandoned his positions on Aqqaqir Ridge. 2nd Armoured Brigade advanced 4000 yards beyond before encountering the German rearguard, containing both tanks and anti-tank guns, deployed on a ridge near Tel el

Mampsra. The usual duel began, with losses on both sides. As the rearguard began to give ground a single PzKw III unexpectedly crossed the crest of a dune nearby and was promptly set ablaze. For a moment the crew lay pinned down in the sand, then a tall figure stood up and walked slowly forward, carrying a coat over his left arm, together with a small canvas bag containing his possessions, his red and gold gorgets and gold shoulder straps revealing that he was a general officer. Captain Grant Singer, who had given up a comfortable post as an ADC to command the 10th Hussars' Reconnaissance Troop during the battle, drove forward in his Daimler scout car to pick him up. As the two exchanged salutes the captive informed Singer that he was Lieutenant General Ritter von Thoma, commander of the Deutsches Afrika Korps. Although the evidence was clear for all to see, Rommel had refused to believe reports that the Eighth Army had broken through, and he had come forward to verify the situation for himself. Singer drove the general to Army Headquarters, where Montgomery gave him dinner, then returned to his troop. The following day, while leading his Regiment's continued advance, he was killed by an Eighty-Eight. In one of those chivalrous gestures which sometimes set the North African Campaign apart from the ruthlessness of total war, von Thoma wrote a considerate note to his parents; this is now preserved in the museum of the King's Royal Hussars.

On 5 November Custance's 8th Armoured Brigade, swinging right towards the coast, ambushed and destroyed a column of 29 Italian and fourteen 14 German tanks, four guns and 100 lorries at Galal, taking 1000 prisoners. This was apparently all that remained of the Italian XX Corps.

The Second Battle of Alamein was over. The cost to the Eighth Army had been 13,500 casualties, approximately 500 tanks knocked out, of which all but 150 were repairable, and 110 guns destroyed by shellfire. Rommel, on the other hand, had sustained 55,000 casualties, including 30,000 prisoners, and his material losses included 450 tanks, 1000 guns and 84 aircraft. His Italian divisions, lacking transport, had been destroyed where they stood and his German divisions, reduced to skeletons, were heading west as fast as they could. In the immediate aftermath of his defeat he was to receive a further stunning shock with the news that the Anglo-American First Army had landed in Morocco and Algeria on 8 November. The war in North Africa had just six months to run.

War to the Death
Tank Battles of the Russian Front 1943/44

R ussian winters provided an experience no tank crewman was ever likely to forget, for apart from creating acute problems of personal survival the unbelievable cold, deepened by the slicing wind, affected their vehicles in strange and unexpected ways. Track and firing pins became brittle and snapped; unless it was of Arctic quality, the hydraulic fluid froze in main armament recoil buffers, rendering the weapon useless; ammunition refused to fire; oil congealed to the consistency of treacle and at first the Germans lacked anti-freeze for their cooling systems so that when the tanks were in leaguer slow fires had to be kept burning beneath their sumps and the engines started every 30 minutes. Later in the campaign, one German armoured regiment found itself temporarily immobilised when a horde of harvest mice, seeking warmth and shelter, penetrated the engine compartments and made a meal of the electrical insulation. Touching bare metal with an ungloved hand resulted in severe burns and loss of skin. It was, of course, as bad for the Russians, but they were at least dressed for the climate and their tanks had been designed for it; the Germans, in sincere admiration of the T34's performance, nicknamed it The Snow King.

The 1942 campaign began in April with an ill conceived Soviet attempt to recapture Kharkov. It ended disastrously with the loss of 1200 tanks and over 250,000 men taken prisoner, proving that the Red Army had not yet reached the state of training at which it could hope to match the Germans in open combat and hope to win. Almost immediately the Germans, still in the throes of their re-equipment programme, opened a major offensive intended to secure the Caucasus oilfields. During this Hitler developed as great a fixation on the capture of Stalingrad (Volgagrad), as he already had on that of Leningrad (St Petersburg), and began committing troops and resources to the battle which was raging for possession of the city. The Russians, who had wisely foregone the temptation to indulge in a second major tank battle like Brody-Dubno, interpreted the situation correctly and prepared a devastating counter-stroke. On 19 November the South West and Stalingrad Fronts routed the Romanian armies holding the line to the north and south of Stalingrad and converged to trap the German Sixth Army within a pocket, together with part of the Fourth Panzer Army. Goering was unable to fulfil his promise to supply the besieged troops by air and an attempt to break through to them was fought to a standstill. On 2 February 1943 the newly created Field Marshal von Paulus, commander of the Stalingrad pocket, surrendered; 120,000 of his men had already been killed in the fighting and of the 91,000 who marched into captivity only a handful were to return home.

In the meantime, to avoid being encircled in their turn, the German armies in the Caucasus had withdrawn. There was, however, a real danger that Army Group South, now commanded by Field Marshal Erich von Manstein, would be trapped by two Soviet Fronts, Voronezh under Golikov and South-West under Vatutin, which were driving west towards the Dnepr, with the ultimate aim of swinging south to the Black Sea coast. Manstein, regarded by many as the best strategist of the war, had only 350 tanks at his disposal, but he planned to use these in a concentrated thrust into the enemy's southern flank, giving him a local superiority of seven to one. For once, Hitler, chastened by the catastrophe at Stalingrad, did not interfere. The counter-offensive commenced on 20 February and immediately found many of Vatutin's columns stalled for want of fuel and short of ammunition. By the end of the month the remnants of South-West Front had been thrown back across the Donets, leaving behind 615 tanks,

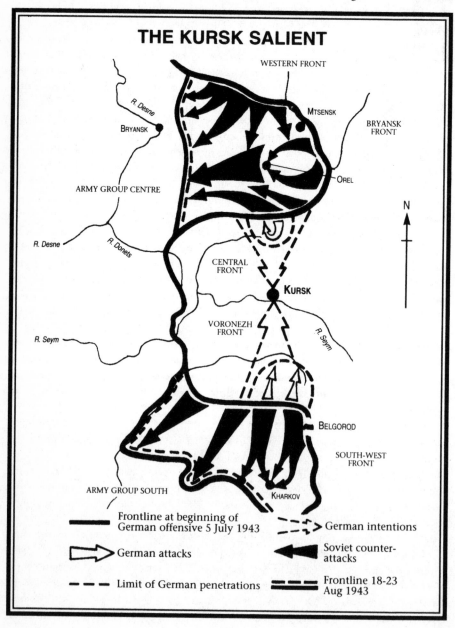

THE KURSK SALIENT

WESTERN FRONT

R. Desne

BRYANSK

MTSENSK

BRYANSK
FRONT

ARMY GROUP CENTRE

OREL

R. Desne

R. Donets

N

CENTRAL
FRONT

KURSK

VORONEZH
FRONT

R. Seym

R. Seym

BELGOROD

SOUTH-WEST
FRONT

ARMY GROUP SOUTH

KHARKOV

— Frontline at beginning of German offensive 5 July 1943

--→ German intentions

⇨ German attacks

◄ Soviet counter-attacks

--- Limit of German penetrations

Frontline 18-23 Aug 1943

400 guns, 23,000 dead and 9000 prisoners. Next it was the turn of Golikov, who was being simultaneously counter-attacked from the west; he too was routed with the loss of 600 tanks, 500 guns and 40,000 men. The spring thaw, turning the roads into bottomless quagmires, brought the pursuit to an end, but by 25 March a new German front had been established.

It was during this series of operations that the Tiger E, which had already been committed to action on the Leningrad sector and in Tunisia, first made its presence felt in southern Russia, the course of the action being recorded by the Halder team:

'Normally, the Russian T34s would stand in ambush at the hitherto safe distance of 1350 yards and wait for the German tanks to expose themselves upon their exit from a village. They would then take the German tanks under fire while they were still outranged. Until now, these tactics had been foolproof. This time, however, the Russians had miscalculated. Instead of leaving the village the two Tigers (from Heavy Tank Battalion 503) took up well camouflaged positions and made full use of their longer range. Within a short time they knocked out sixteen T34s which were sitting in open terrain and, when the others turned about, the Tigers pursued the fleeing Russians and destroyed eighteen more tanks. It was observed that the 88mm armour piercing shells had such a terrific impact that they ripped off the turrets of many T34s and hurled them several yards. The German soldiers' immediate reaction was to coin the phrase, "The T34 raises its hat whenever it meets a Tiger!" The performance of the new German tanks was a great morale booster.'

The ending of Manstein's counter-offensive had left a huge salient in the line, one hundred miles wide and seventy deep, centred on the city of Kursk. As the North African campaign drew towards an end that was as disastrous for the Third Reich as Stalingrad, Hitler sensed that tide of war had turned against him and reached the conclusion that by trapping and destroying a large part of the Soviet Army within the Kursk salient he could induce Stalin to conclude a negotiated peace, thereby leaving him free to confront his enemies in the west. However, whereas the plans for Manstein's counter-offensive had been worked out in the secure headquarters of Army Group South, those for the attack on the Kursk salient, codenamed 'Citadel', were made at OKW, which had been penetrated by the Lucy spy ring, so that every detail was quickly passed to the Russians. The latter therefore began turning the salient into a huge fortress, girdled by three defensive belts to a depth of 25 miles, covered by 20,000 guns of which one-third were anti-tank weapons, and surrounded by minefields laid to a density of 2500 anti-personnel and 2200 anti-tank mines per mile of front.

As finalised, the plan for Citadel involved converging thrusts through the northern and southern shoulders of the salient, which would then become a giant pocket. Some German commanders believed that the Tiger and the new Panther would guarantee victory. Others had serious reservations about the idea. Some, correctly, believed that the objective was too obvious and that the Russians were already preparing defences in depth. Others did not wish to see the entire potential of the carefully reconstructed armoured corps committed to what they saw as a gamble. Guderian, now Inspector General of Armoured Troops, made it clear that the Panther's teething troubles were far from over and that, whatever the outcome, tank losses were bound to be heavy. Nevertheless, the concept received sufficient support for it to be approved. The stage was now set for the greatest tank battle in history, in which 900,000 German and allied troops with 2700 fighting vehicles confronted 1,337,000 Russians with 3300 tanks and assault guns.

Even before the battle began on 5 July Guderian's worst fears were realised. Given a little more time, the Panther would be developed into the best medium tank of the war but, for the moment, it was simply not ready for action. The tracks between rail-heads and assembly areas became littered with breakdowns that were mainly the result of transmission failures and engine fires. On the sector of Hoth's Fourth Panzer Army more breakdowns and battle casualties reduced the theoretical number of available Panther's from 200 to just forty on the first day of the battle.

After a week's bitter attritional fighting the northern wing of the attack had advanced a mere ten miles. The southern wing had done a little better, penetrating some

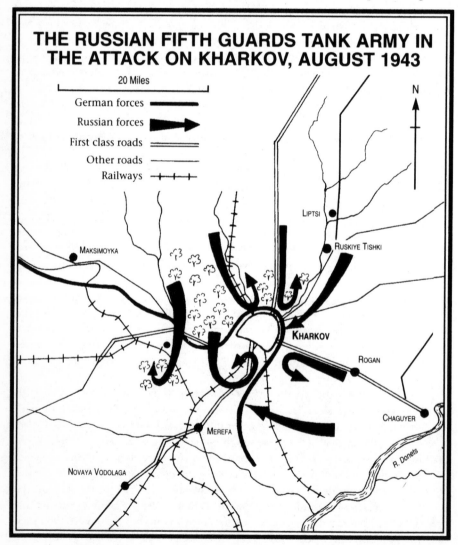

THE RUSSIAN FIFTH GUARDS TANK ARMY IN THE ATTACK ON KHARKOV, AUGUST 1943

20 Miles

German forces
Russian forces
First class roads
Other roads
Railways

N

LIPTSI

MAKSIMOYKA

RUSKIYE TISHKI

KHARKOV

ROGAN

CHAGUYER

MEREFA

R. Donets

NOVAYA VODOLAGA

25 miles, and it was in this area that the climax of the battle was reached on 12 July when Hoth's 700 tanks met the 800 of General P. A. Rotmistrov's Fifth Guards Tank Army in a headlong clash that developed into a gigantic mêlée in which there was little or no place for tactical finesse.

It began at about 0800 with the Germans advancing north-eastwards towards Prokhorovka with the morning sun in their eyes. Suddenly waves of T34s appeared, approaching at speed obliquely from the right, and without pausing they drove straight through the leading German ranks. Such extreme action, never witnessed before or since on any battlefield, ran quite contrary to the tactical manual, but the Russians had their reasons and these are set out in their official history.

'It destroyed the enemy's ability to control his leading units and subunits. The close combat deprived the Tigers of the advantages which their powerful gun and thick armour conferred, and they were successfully shot up at close range by the T34s. Immense numbers of tanks were mixed up all over the battlefield; there was neither time nor space to disengage and reform ranks. Fired at short range, shells penetrated front and side armour. There were frequent explosions as ammunition blew up, throwing tank turrets dozens of yards from their stricken vehicles....On the scorched black earth, smashed tanks were blazing like torches. It was difficult to tell who was attacking and who was defending. The fortunes of the combatants varied from place to place.'

Both sides committed more troops to the battle. Ramming became commonplace; one burning KV, trailing flames and smoke and manned only by its driver, smashed headlong into the Tiger that had inflicted the fatal wounds and in the shattering explosion which ensued both tanks were destroyed. Overhead, ferocious dog-fights continued throughout the day as both air forces strove to assist their ground troops.

'The landscape,' Rotmistrov recalled, 'Seemed too small for what was taking place upon it. Smoke and dust darkened the sky. Separate shots could not be heard as all sounds blended into a continuous terrific roar.'

By the time dusk put an end to the fighting the Fourth Panzer Army had lost about 300 tanks, including 70 Tigers, most of which had to be left on the battlefield. The Fifth Guards Tank Army lost some 400 tanks, or half its strength, but it had halted the German drive. This brought each side's tank losses since 5 July to somewhat in excess of 1500 vehicles. For the Russians, who possessed the greater manufacturing capacity and were able to recover most of their casualties, the long term implications were clearly less serious than for their opponents.

After Prokhorovka there was little chance that Citadel would succeed. On 17 July Hitler announced that the assault on the Kursk salient, in which the reforged potential of the German armoured corps had been squandered for little return, was over. From this time onwards the strategic initiative rested with the Russians, and they would retain it for the rest of the war. Yet, in the apparently endless series of defensive battles it was to fight during its long retreat, the German armour continued to demonstrate its tactical superiority, as Rotmistrov's refitted 5th Guards Tank Army found to its cost when it attempted to encircle Kharkov the following month.

'Kharkov constituted a deep German salient to the east, which prevented the enemy from making use of this important traffic and supply centre,' recorded the Halder team. 'When the Russian command perceived its mistake, Stalin ordered the immediate capture of the city and assigned the mission to the 5th Guards Tank Army. The German XI Corps, whose five infantry divisions sealed off the city in a long arc, recognised the new danger in time. It was clear that the Russian tanks would not make a frontal assault on the projecting Kharkov bastion, but would attempt to break through the narrowest part of the arc to the west in order to encircle Kharkov. All available anti-tank guns were set up on the northern edge of the bottleneck, which rose like a bastion. Flak guns were set up in depth on the high ground. The anti-tank defence would not have been sufficient to repulse the expected mass attack of Russian tanks, but at the last moment 2nd SS Panzer Division Das Reich arrived with strong armoured reinforcements and was immediately despatched to the most dangerous sector.

'The 96 Panthers, 32 Tigers and 25 assault guns had hardly taken their assigned positions when 5th Guards Tank Army commenced its first large scale attack. The first hard German blow, however, hit the assembled mass of Russian tanks while they were still assembling in the villages and the flood plains of a brook valley. Escorted by German fighters, which cleared the sky of Russian aircraft within a few minutes, wings of heavily laden Stukas came on in wedge formation and unloaded their cargoes of destruction in well aimed dives on the assembled tanks. Dark fountains of earth erupted skywards, to be followed by thunderclaps and shocks resembling an earthquake. These were the result of two-ton bombs, designed for use against battleships, which were all that the Luftwaffe could lay its hands on to counter the attack. Wing after wing approached with majestic calm and carried out its work of destruction without interference. Soon all the villages occupied by the Soviet tanks were in flames. A sea of dust and smoke clouds illuminated by the setting sun hung over the brook valley. Dark mushrooms of smoke from burning tanks stood out in sharp contrast. The Russians were hit so hard that they were unable to launch their projected attack that day.

'The next day the Russians avoided mass grouping of tanks, crossed the brook valley at several places and disappeared into the broad cornfields which were located ahead of the front but which ended at the east–west highway several hundred yards in front of the main line of resistance. During the night motorised infantry had already infiltrated through the defence lines in several places and near Lyubotin they made a surprise penetration into the artillery position. After stubborn fighting the gun crews were forced to retreat, taking their breech blocks with them, and twelve howitzers fell into the enemy's hands. The leading elements of the enemy infantry were already shooting it out with German defence platoons in a wood adjoining the corps headquarters.

'During the morning Red tanks worked their way forward in the hollows up to the southern edge of the cornfields. Then they made a mass dash across the road in full sight. The leading waves of T34s were caught in the fierce defensive fire of the Panthers and were set ablaze before they could reach the main line of resistance. But wave after wave followed and pushed forward into the battle position. Here they were trapped in a net of anti-tank and anti-aircraft guns, Hornets [heavy tank destroyers consisting of an 88mm gun mounted on a PzKw IV chassis, more often referred to as Rhinos], and Wasps [self-propelled 105mm howitzers on a PzKw II chassis], were split up, and large numbers of them were put out of action. The last waves were still attempting to force a breakthrough in concentrated masses when they were attacked by Tigers and assault guns, until then the mobile reserve behind the front, and were repulsed with heavy losses. The price they paid for this mass tank assault amounted to 184 knocked out T34s.

'In the meantime, German infantry reserves supported by assault guns from the 3rd Panzer Division had recaptured the lost battery position and bottled up the battalion of enemy infantry that had infiltrated the position. Stubbornly defending themselves, the Russians waited for the help that their radio had promised.

'The Russians changed their tactics next day and attacked further east in a single deep wedge, using several hundred tanks simultaneously. But even while they moved across open terrain along the railway, numerous tanks were set on fire at a range of 2000 yards by the long range weapons of the Tigers and Hornets. The attack was not launched until late in the forenoon. As the tanks emerged from the cornfields they were assailed by the concentrated fire of every available Tiger, Hornet, Panther, assault gun, flak and anti-tank gun and in a short time the attack collapsed with the loss of 154 tanks. The weak rifle units which had followed were mown down by the concentrated fire of the German infantry and artillery as they emerged from the cornfields. The encircled Red infantry battalion had waited in vain for assistance, but continued to fight on with incredible tenacity; after 48 hours of heroic defence the battalion was killed to the last man.

'The losses incurred by the Russians were enormous. However, they still possessed more than 100 tanks and experience had taught that further attacks were to be expected. The few tank crewmen taken prisoner were aware that death or, if they were lucky, capture, awaited every one of their comrades.

'Contrary to all expectations, an eerie calm prevailed throughout the following day. Several Red tanks crawled about in the cornfields and towed damaged tanks away in order to reinforce their greatly depleted ranks. Summer heat shimmered over the bloody fields of the past days of battle. A last glow of sunset brought the peaceful day to a close. Might the enemy have given up his plan, or even refused to obey the supreme order to repeat the attack? He came back, and on the same day. Before midnight, considerable noise from tanks in the cornfields betrayed his approach. He intended to achieve by night what he had failed to do by day.

'Before he had reached the foot of the elevated terrain, numerous flashes from firing tanks had ripped the pitch black darkness of the night and illuminated a mass attack by the entire 5th Guards Tank Army on a broad front. Tanks knocked out at close range were already burning like torches and lighting up portions of the battlefield. The German anti-tank guns could no longer fire properly, since they could hardly distinguish between friend and foe; German tanks had entered the fray, ramming Russian tanks in a counterthrust or penetrating them at gun barrel range in order to block the breakthrough. A steady increase in the flash and thunder of tank, anti-tank and flak guns could be perceived after midnight. Farm buildings went up in flames, illuminating the plateau on which this great night tank duel was being fought. This made it possible to identify the Red tanks at a range of over 100 yards and to engage them. The thunderous roll turned into a din like the crescendo of kettledrums as the two main tank forces clashed. Gun flashes from all around ripped the darkness over a wide area. For miles, armour piercing projectiles whizzed into the night in all directions. Flashes also appeared further and further behind the lines but after two or three hours of pandemonium the front was restored.

'After daybreak the Germans sensed that the battle was won, although there were still Red tanks in and behind their position, and here and there a small gap remained to be closed. Mopping up lasted all morning. Only a small patch of woodland, close behind the main line of resistance, was still occupied by Red motorised infantry supported by a few tanks and anti-tank guns. All attempts to retake this failed with heavy losses; not even concentrated artillery fire could force the Russians to yield. This tenacious resistance was ended only by an attack of flame throwing tanks which burned the entire strip of woodland to the ground.'

In three days of heavy fighting Rotmistrov's army, so highly regarded as a tactical élite that its members were known as The Professors, had lost 420 tanks and suffered such heavy losses of men and equipment that, for a while, it ceased to be of any fighting value. Nevertheless, because of Soviet pressure elsewhere, the front lines moved inexorably westward and the Germans had to abandon Kharkov.

As the Soviet advance continued, the German armour attempted to counter mass with mobility and tactical flexibility. Russian attacks still relied heavily on preplanning and rehearsals, each success being purchased at a terrible cost that would not have been tolerated in any western army. The predictability of Soviet tactics ensured that the Germans were often able to plan their own response in advance. Thus, while the weight of an attack might be absorbed frontally, sometimes by evacuating the forward positions, upon which the formidable Soviet artillery would continue to waste ammunition, the German armour could be used to strike into the attackers' flanks or rear at the critical moment, or even deep into Russians' rear areas, so destroying the enemy's entire operational plan. The Germans, however, could not be strong everywhere and the early part of 1944 was marked by a series of engagements known as Cauldron Battles, involving formations that had been bypassed by the Russian advance and later surrounded. Sometimes, as at Cherkassy, the encircled troops, with external assistance, were able to break out at the cost of much of their equipment. An alternative to this was provided by General Hans Hube's First Panzer Army, which became a 'wandering pocket' and fought its way westwards spearheaded by its Panther battalions, its rear covered by Tigers until it reached its own lines; on the way, it destroyed no fewer than 357 Soviet tanks and assault guns.

During this period the Tiger battalions were employed in the fire brigade role, moving from one threatened sector to another, winning their battles but too few in number to influence the course of the campaign. One such unit, named Panzer Regiment Bäke after its commander, Lieutenant Colonel Dr Franz Bäke, a veteran of the tank battle at Cambrai in October 1918, is especially worthy of mention. Formed in January 1944, the regiment initially consisted of Heavy Tank Battalion 503 with 34 Tigers, and II/Panzer Regiment 23 with 47 Panthers, a self-propelled artillery battalion and an engineer bridging battalion. The unit's tremendous firepower was dramatically demonstrated when it was first committed to action on 26 January in an attempt to halt the advance on Vinnitsa of five Soviet tank corps. During five days and nights of heavy fighting it destroyed 267 tanks and 156 guns of various types at the cost of one Tiger and four Panthers.

The Tiger's reign, however, was coming to an end. One day in May, near Targul Frumos in Romania, the Panzergrenadier Division Grossdeutschland's heavy tank battalion opened fire at 3000 yards on the unfamiliar shapes of Soviet IS-IIs. Hitherto, the Tiger commanders and gunners had confidently observed the flat trace of their armour piercing rounds streak towards the target, seen the flash of the strike and watched their opponents grind to a halt, belching smoke and flames. On this occasion, however, the rounds simply flew off the enemy's thick frontal castings without doing the slightest damage. Not until the range had closed to less that 2000 yards did the Tiger crews began to score kills, discovering to their horror they were simultaneously vulnerable to the enemy's 122mm gun.

By June 1944 the scales had tilted irrevocably against the German armoured corps. The concurrent demands of the Russian Front, Italy and now Normandy meant that it was constantly overstretched. It had always fought in close partnership with the Luftwaffe, but the British and American air offensive against Germany meant that most of the latter's fighter aircraft were withdrawn for the defence of the homeland, and without them the Stuka dive bombers were sitting ducks; for the same reasons, thousands of 88mm flak guns were retained in Germany at a time when their tank killing capacity was most needed in the east. In 1944 Germany's tank and self-propelled gun production reached a peak of 19,000 vehicles, but there were never enough to go round. The armoured regiments of the panzer divisions, consisting of one battalion equipped with Panthers and one with upgunned PzKw IVs, were permanently under strength and increasingly forced to supplement their strength with assault guns and tank destroyers. In sharp contrast, the USSR alone produced 29,000 tanks and self-propelled guns the same year, including 11,000 T34/85s and 2000 IS-IIs; in addition, the Soviet Army also received supplies of Churchills and Shermans from the Western Allies. For the Germans, therefore, it seemed that no matter how many Soviet tanks they destroyed, there were always more to follow.

During Operation 'Bagration', which began on 22 June, three Soviet fronts overwhelmed Army Group Centre. No less than 40 tank brigades and numerous cavalry mechanised groups roared across eastern Poland, displaying a newly acquired expertise as they encircled one objective after another. Somehow, as the offensive ran down, the line was re-established, but ground lost meant that Army Group North was all but isolated with due course it had to be evacuated by sea. From this point onwards it was obvious that the Third Reich could not survive for long. Most German soldiers continued to fight hard in the hope that by keeping the Russians out of their homeland for as long as possible the Western Allies would reach it first. Robert Poengsen, the German war correspondent, describes a typical action fought by 4th Panzer Division during these desperate times.

'July 1944 – in the evening the 3rd Company of Panzer Regiment 35 and the men of Panzergrenadier Regiment 12 reach the village. Although the situation is secure enough for the present, the enemy advance has driven a deep wedge through our front and we may have to fight our way out. Over the radio the battlegroup is ordered to move.

'Far behind our artillery has already moved to new positions. The armoured personnel carriers roll forward. One route remains open. The Panthers cover the retreat of their comrades. They remain in the village and keep the enemy in check with their long guns. The order comes to disengage.

'Lieutenant Goldhammer, the company commander, declares that the situation has become extremely serious. The route taken by the Panzergrenadiers is unsuitable for tanks, because of a bridge that will not support their weight. Only one alternative remains - we must fight our way through the Russians.

'Mount! Start up! Advance! The tanks move out and form line-ahead while shells burst here and there in the village, which soon begins to burn. The hatches are closed, the guns ready for action. The crews wait expectantly for the command, anxious about what lies ahead.

'Lieutenant Gohrum's tank, in the lead, reaches the exit from the village. Heavy fire is directed at him. Speed up! Let's get out of here! The driver presses his accelerator. 30, 40, 45 kilometres per hour reads the speedometer. From the right a Sherman opens fire. Gohrum halts quickly. Two, three shots and flames belch out of the Russian tank. The driver moves off again. The road becomes wider. Russian infantry open a useless fire with their rifles from a wood. We are out of the village. Let's go – speed up! A wild hunt begins. Panic-stricken earth-brown figures leap from the road in front and vanish into the cornfields; however, the enemy also possesses anti-tank guns. The tank engines race. Long muzzle flashes come from the guns. House-high dust plumes rise from the grinding tracks. All turrets are traversed right to 3 o'clock, where gun flashes can be seen in the cornfields. Wherever commanders and gunners look there are anti-tank and artillery guns firing. They feel the impact of hits and fire back on the move with all weapons. To halt is suicide. Only Luck can help. It is a drive through Hell. To turn aside is impossible. Everywhere there are explosions and flashes. There are glaring jets of flame whenever a shell bursts against the tanks' angled armour.

'In their foxholes the Panzergrenadiers are aware that the Russian fire has risen to a climax. Anti-tank rounds whistle overhead. Suddenly a tank, trailing a huge dust cloud, clatters towards their position, then a second and third, a whole troop. From behind come others, closed up tight together. Tanks to the front! The alarm call is passed from foxhole to foxhole. A lieutenant with binoculars identifies the tanks as German. He jumps up and waves his cap. The first tank steers towards him. The turret opens and light shines upwards, revealing a black forage cap. The tanks halt a few metres in front of our grenadiers, one behind the other. The hatches are flung open. The crews blink in the last light of day. They clamber out and breath in the fresh air after their incredible journey. While they smoke a cigarette they examine hits on the turrets and hulls. They look back towards the distant village, now in flames, and are glad to be alive.'

Hobo's Funnies

When, on 19 August 1942, the Allies mounted a major raid on Dieppe, they did so with the political object of demonstrating to Stalin that the opening of a 'Second Front' in the West was, for the moment, not a practical proposition. However, while the operation was ostensibly a failure, many valuable lessons were learned and put to good use in the planning of the invasion of Western Europe almost two years later.

After Dieppe the Allies recognised the impossibility of securing a heavily defended French port, deciding instead to invade over the open beaches of Normandy and bring their own prefabricated harbour with them. These beaches, while not as heavily fortified as the more important ports, nonetheless possessed formidable defences which were capable of inflicting terrible losses on a conventional landing force. The German belief was that the Allies would time their invasion to coincide with high tide, so reducing the period that their troops would be dangerously exposed while crossing the beach. They therefore constructed lines of obstacles that would remain submerged for a considerable period either side of high water. These included rows of fixed stakes, hedgehogs and tetrahedra made from girders, and triangular constructions known to the Allies as Element C, to all of which explosive devices had been attached. Such obstructions not only presented a physical barrier but were also capable of disembowelling any landing craft which tried to batter its way through.

The beaches themselves were heavily mined and covered by the fire of machine-guns and artillery weapons housed in concrete bunkers. Where no sea wall existed, high concrete anti-tank walls had been built to block natural exits from the shore. Behind, there were strongpoints containing more guns and automatic weapons, protected by mines, wire entanglements and anti-tank ditches, laid out in depth. Houses and other buildings, specially strengthened, had also been brought into the defensive scheme.

The difficulties facing the Allies were therefore considerable but not insuperable. It was decided that the landings would be made at half tide, when the water was beginning to rise but the lines of obstacles were still exposed. These would be tackled by underwater demolition teams of naval frogmen, who would first disarm the enemy's explosive devices then clear 50-yard gaps for the passage of the landing craft. Beyond this point, the reduction of the defences became the responsibility of the ground troops.

In the British Army it was generally recognised that the key to the various problems associated with assaulting such long-prepared defensive positions lay in the use of armoured vehicles, just as it had in World War I. In April 1943 Major General P. C. S. Hobart, Commander of 79th Armoured Division, which had been raised the previous October as a conventional armoured formation, was informed that the Army's specialist tank and assault engineer units would be concentrated under his command with the object of developing techniques and equipment that would be used to spearhead the coming invasion. At the time few of the division's personnel could have imagined that they were destined to occupy a unique place in the annals of armoured warfare.

Hobart was unquestionably the man for the job. He had begun his career with the Bengal Sappers and Miners and was therefore familiar with assault engineering methods. After World War I he had transferred to the Royal Tank Corps and had commanded the 1st Tank Brigade in 1934. However, as he rose in rank, his innovations and bluntly expressed ideas on the mechanisation of the Army began to make him unpopular with the War Office; nor did it help that he was generally proved right. The problem was that while his insight amounted to something like genius, he had no patience with those who were not similarly gifted. Sometimes, members of his staff would be greatly touched by some personal act of kindness, but most regarded the experience of serving under him as being 'absolute hell'; those officers who did not instantly and fully comprehend what the General was talking about were left wishing they had never been born. This might have been all very well in its way had not Hobart, who was no respecter of rank, dealt similarly with his superiors. Warned repeatedly by his friends to moderate his behaviour, he expressed genuine surprise that feelings should have been ruffled, and carried on as usual. Known as a fine trainer, in 1938 he was sent to Egypt to form and train the Mobile Division, as the 7th Armoured Division was initially known, and brought it to an outstanding pitch of efficiency and desert worthiness. Unfortunately, he was not to see the results of his work, for following yet another personality clash, this time with the GOC Egypt, he was removed from command and returned home. Early in 1940 he retired from the Army and joined the Home Guard, where his drive and energy quickly earned him promotion to lance corporal. At this point Winston Churchill, recognising the absurdity of losing one of the Royal Armoured Corps' most outstanding officers at a time of national emergency, recalled him to form and train in succession the 11th and 79th Armoured Divisions.

In Hobart's opinion not even specialist armour could be expected to produce the required results unless it received direct gunfire support from other tanks. This meant that conventional tanks would have to be landed in the first assault wave, before the beach defences had been neutralised by the specialist armour, a conundrum of apparently chicken-and-egg proportions to which the answer was provided by the DD (Duplex Drive) 'swimming' tank. The concept of such an amphibian, kept afloat by a collapsible canvas screen attached to the hull and driven by a screw which drew its power from the main engine, had been pioneered by Mr Nicholas Straussler during the inter-war years; the idea being that once the vehicle had reached the shoreline the buoyancy screen would be lowered and the normal drive engaged, enabling it to perform as a normal gun tank. Successful trials had already been carried out with the Valentine and the Stuart, but the choice for the Normandy landings fell on the Sherman because of its superior firepower. Hobart envisaged that, initially at least, the DDs would provide fire support from the shallows until the specialist armour had produced sufficient elbow room, then fight as the situation warranted. Among those units trained by 79th Armoured Division in the use of the DD were the 4th/7th Dragoon Guards, the 13th/18th Hussars, the 15th/19th Hussars, the Sherwood Rangers Yeomanry, the East Riding Yeomanry, the Canadian 1st Hussars and Fort Garry Horse, and the American 70th, 741st and 743rd Tank Battalions. Various methods of mechanical mineclearing existed, but experience at Alamein had shown that the best results were obtained by flailing, that is beating the ground ahead of a tank with chains attached to a rotating drum powered by the vehicle's main or auxiliary engine, so detonating any mines in its path. The most successful design was the Sherman Crab, capable of clearing a lane almost ten feet wide at a speed of 1.25 miles per hour; when it was not flailing the Crab could fight as a normal gun tank. The 30th Armoured Brigade, consisting of the 22nd Dragoons, 1st Lothians and Border Horse and the Westminster Dragoons, joined 79th Armoured Division in November 1943 and was immediately equipped with Crabs, which it retained for the rest of the war.

The most versatile vehicle in the division's armoury was the AVRE (Assault Vehicle Royal Engineers), which had been developed as a direct result of the experience gained in the Dieppe raid. The hull of the Churchill infantry tank was selected as the basis because of its heavy armour, roomy interior and obvious adaptability. The AVRE was fitted with a specially designed turret mounting a 290mm muzzle loading demolition gun, known as a petard, which could throw a 40-pound bomb, designed to crack open concrete fortifications, to a maximum range of 230 yards; reloading was carried out through a sliding hatch above the co-driver's seat. Standardised external fittings enabled the vehicle to be used in a variety of ways. It could carry a chestnut paling (chespale) fascine, up to eight feet in diameter and fourteen feet wide, that could be dropped into anti-tank ditches, forming a causeway; it could lay an Small Box Girder (SBG) bridge with a 40-ton capacity across gaps of up to 30 feet; it could be fitted with a Bobbin which unrolled a carpet of hessian and metal tubing ahead of the vehicle, so creating a firm track across soft going; it could place 'Onion' or 'Goat' demolition charges against an obstacle or fortification and fire them by remote control after it had reversed away; it could be used to push Mobile Bailey or Skid Bailey bridges into position; and, on going which was not suited to the Crab, it could be fitted with a plough which brought mines to the surface. The 1st Assault Brigade RE, consisting of the 5th, 6th and 42nd Assault Regiments RE (ARRE), each with an establishment of 60 AVREs and a number of D8 armoured tractors, was formed as part of the division during the summer of 1943.

Naturally, the nature of its work meant that Hobart's division had to function along the lines of a large plant hire organisation. No two sectors of the German coast defences were exactly alike, and each presented the attacker with its own set of problems. Liaison officers were therefore attached to each of the British and Canadian infantry divisions forming the first wave of Lieutenant General Sir Miles Dempsey's British Second Army, which formed the left wing of the Allied invasion force. After discussions involving the detailed study of maps, models, the most recent air reconnaissance photographs and intelligence reports, the appropriate assault engineering equipment was allocated and formed into teams which received a similarly detailed briefing on their own specific tasks. These facilities were also offered to Lieutenant General Omar Bradley, commander of the US First Army, which formed the right wing of the Allied assault; apart from the DD battalions mentioned above, they were declined for reasons that have never been satisfactorily explained, with tragic consequences.

On the morning of D Day itself, 6 June 1944, sea conditions were never less than difficult and the fortunes of the DD units varied considerably. Off Utah Beach, the US 70th Tank Battalion launched 30 tanks from its LCTs 3000 yards out and all but one reached the shore safely. In contrast, off Omaha Beach the US 741st Tank Battalion launched 29 tanks 6000 yards out and all but two were swamped and sank. Off Juno, A Squadron of the Canadian 1st Hussars launched ten tanks at between 1500 and

2000 yards, of which seven touched down; the regiment's B Squadron launched nineteen from 4000 yards, of which fourteen touched down. The day's most successful performance was put up by the 13th/18th Hussars who launched 34 tanks some 5000 yards off Sword Beach and landed with 31. Elsewhere, conditions were so bad that launching was never contemplated. The US 743rd Tank Battalion landed on Omaha direct from its LCTs; as did, in the British sector, the 4th/7th Dragoon Guards and Sherwood Rangers Yeomanry on Gold and the Fort Garry Horse on Juno, arriving after the armoured assault teams.

The German infantry divisions manning the fortifications of Hitler's Atlantic Wall contained a high proportion of non-Germans and were regarded by some as being second line troops. Despite this, and the fact that they had been forced to endure air attacks and a mind-numbing naval bombardment, they were prepared to offer the most stubborn and tenacious resistance. They were, however, quite unprepared for the DDs and the wave of specialist armour which was now being disgorged from its LCTs onto the beach, for so good had been the pre-invasion security that the existence of such devices had never been suspected.

For example, while swimming the DD revealed only a few inches of its floatation screen and from a distance resembled a harmless ship's whaler. Once the vehicle emerged from the shallows, however, the screen dropped to reveal a Sherman spitting fire. On Juno Beach the Canadian infantry were pinned down until A Squadron 1st Hussars came ashore under heavy mortar and shellfire and quickly eliminated strongpoints containing two 75mm guns, one 50mm gun and six machine-guns, after which a large party of the enemy came forward to surrender.

Even the most meticulous of military planners works on the assumption that the natural state of war is chaos, and he allows for as many contingencies as he can. It was, therefore, never envisaged that all of 79th Armoured Division's breaching teams would be able to clear lanes through the defences, but a sufficient margin of safety had been left for the momentum of the attack to be maintained and enable the assault divisions to start moving inland. It is impossible in these few pages to describe the actions of all the division's teams, but the following are representative.

On Nan Sector of Juno Beach, where the 8th Canadian Infantry Brigade was to come ashore between Bernières-sur-Mer and St Aubin-sur-Mer, the breaching teams consisted of Crabs manned by B Squadron 22nd Dragoons and the AVREs of 80 Assault Squadron RE (5 ARRE). The DDs had yet to arrive and the teams' LCTs, also behind schedule, touched down on the rising tide. With the area of exposed beach shrinking steadily and becoming crowded with infantry, No 1 Team's leading Crab flailed up to the sea wall against which an SBG AVRE laid its bridge. Unfortunately, the first AVRE to surmount this struck a mine a little way further on and was immobilised, blocking the intended exit. Nearby, however, a section of the sea wall had been partially blown down by the preparatory bombardment and two Crabs not only succeeded in flailing their way up to the gap but also scrambled over it and reached the lateral road beyond, which they proceeded to clear. Two fascine AVREs followed and dropped their bundles into the anti-tank ditch beyond. Later, the damaged AVRE was pushed aside by a bulldozer, the driver of which was almost immediately killed by a mine, and the gap through the minefield was completed by hand, creating a second exit.

Because of congestion in the approaches to the beach, No 2 Team's LCTs touched down some 300 yards east of their intended target. As the AVREs emerged they were immediately engaged by a 50mm anti-tank gun firing from the west. The SBG AVRE was knocked out and another AVRE commander was killed before the remainder silenced the emplacement with their petards. The sea wall, 12 feet high, was only 50 yards away and the Crabs flailed a lane up to it. At this point the loss of the SBG was keenly felt, for although the AVREs damaged the wall with their petards in an attempt to bring it down, they failed to make a wide enough gap and the crater created was too soft and steep for the passage of vehicles. By now the infantry had worked their way forward and drew the team commander's attention to a beach ramp blocked by Element C, which was blown apart by petard fire. The Crabs then flailed the ramp and an AVRE dropped a fascine into the anti-tank ditch, completing the exit.

No 3 Team had a much easier time, despite the fact that one of its LCTs was hit and barely managed to reach the shore. The Crabs flailed a lane to the sea wall, the SBG bridge was positioned and the Crabs crossed it to continue flailing a route through the dunes as far as the lateral road. No 4 Team, on the other hand, was dogged by bad luck. The LCTs came in 150 yards east of their objective, close to the high tide line and in an area where the depth of water varied considerably. An incoming landing craft collided with an SBG AVRE as it disembarked, forcing the crew of the latter to abandon the vehicle; so close were they to the enemy that three were killed and one wounded by sniper fire and grenades. The team's Crabs turned west and flailed a lane to No 3 Team's exit. An AVRE also unrolled its bobbin carpet over an area of soft sand but this was quickly torn apart by the passage of tracked vehicles.

Pending the arrival of the DDs, the infantry had to rely on the Crabs and AVREs to assist them in dealing with those of the defences which were still holding out. 'A pillbox on the cliff fell to petard fire and houses belching forth streams of mortar bombs and small arms fire were silenced by 75mm and petard fire,' recorded the divisional historian. As the accuracy of the petard declined beyond 80 yards, most of these engagements took place at close quarters. The bomb itself, known as The Flying Dustbin, was visible throughout its flight and its effect was devastating, bringing whole sections of house down in a thunder of collapsing brickwork and splintering timber; needless to say, very few hits were required to bring the defenders out into the open.

To the east, the breaching teams on Queen Sector of Sword Beach, where the British 8th Infantry Brigade was coming ashore at Lion-sur-Mer, consisted of A Squadron 22nd Dragoons and 77 and 79 Assault Squadrons RE (5 ARRE). On the right No 1 Team beached at a point overlooked by high sand dunes. In the face of fierce fire, the Crabs flailed up the beach and over the dunes, one commander killing two snipers with a grenade thrown from the turret. The leading AVRE, commanded by Sergeant Kilvert, was hit as it emerged from the LCT and drowned in the shallows. Undeterred, Kilvert and his crew grabbed their personal weapons and made their way across the beach to storm a fortified farmhouse and rout an enemy patrol, later handing over their prisoners to the infantry. The team's remaining AVREs assisted the Crabs in completing a route inland then set off to assist 48 Royal Marine Commando in the capture of Lion.

No 2 Team lay off the beach until the DDs of 13th/18th Hussars had touched down, and in the process drifted west of No 1 Team. The first Crab to disembark, commanded by Sergeant Smyth, immediately charged and crushed the 75mm anti-tank gun that had opened fire on No 1 Team. The Crabs then cleared a lane across the beach until one blew a track on a mine. It was bypassed and an SBG bridge was dropped across the wrecked gunpit, completing the exit. After this, the team's AVREs also headed for Lion.

No 3 Team's Crabs completed one lane, along which a Bobbin AVRE unrolled its carpet; the vehicle then struck a mine and, having also been hit by anti-tank fire, was drowned by the rising tide. A second lane was then flailed, at the end of which an SBG bridge was laid to provide an exit from the beach. No 4 Team's LCT became the target of a heavy calibre gun and was hit repeatedly. The leading Crab got ashore safely but the second was hit while on the ramp and nothing could get past. When more hits caused explosions aboard the craft, killing the sector's senior engineer officer, it was forced to withdraw and sail back to England. When the team's solitary Crab had part of its jib shot away by anti-tank fire, its commander, Lieutenant R. S. Robertson, jettisoned the rest and fought as a gun tank.

As more troops and their supporting armour arrived the battle moved inland while the Crabs continued to flail the beach and the AVREs set about the task of recovering vehicle casualties and assisted in removing beach obstacles. The 79th Armoured Division's breaching teams had employed 50 Crabs and 120 AVREs, of which 12 and 22 respectively had been knocked out; casualties amounted to a total of 169 killed, wounded and missing, which, given the nature of the task which had been set, was astonishing. By midnight on 6 June, 57,000 American and 75,000 British and Canadian troops had been put ashore; and on the British sector alone 950 fighting vehicles, 5000 wheeled vehicles, 240 field guns, 280 anti-tank guns and 4000 tons of stores had been landed.

If, superficially, this suggests an absurdly cheap victory, it must be compared with what happened on the American sectors where, it will be recalled, the troops did not have the benefit of armoured breaching teams. On Utah Beach their task had been eased somewhat by the dropping of the US 82nd and 101st Airborne Divisions some miles inland during the previous night; but on Omaha Beach the US 1st and 29th Infantry Divisions were pinned down on the shore by defences that were no more formidable than those on the British sector. Not until an hour after the initial landings did the situation begin to improve slowly when eight American and three British destroyers, observing the carnage amid the shattered LCIs at the water's edge, closed in to batter the defences at point blank range. Even then, it was only by the inspired leadership, heroism and self-sacrifice of individuals and small groups that the Americans began to make ground little by little. By midnight, while the other beachheads were between seven and nine miles deep, that at Omaha amounted to a foothold extending at best some 2000 yards from the shoreline. The cost had been 3000 casualties, half the American losses for the day, including that of the two airborne divisions; British and Canadian casualties at Gold, Juno and Sword Beaches amounted 4200 killed, wounded and missing.

The events of D Day demonstrated beyond any possible doubt the validity of Hobart's tactical concepts and his thorough training methods. For the remainder of the campaign in North-West Europe elements of 79th Armoured Division, now known throughout 21st Army Group as 'Hobo's Funnies,' played a vital part in every major and countless minor operations fought by the British and Canadian armies, as well as providing support for the Americans when requested.

Enough has been written on the conduct of operations in Normandy for only the briefest of reminders to be given here. Although the terrain, much of which consisted of small fields bounded by hedges set on earth banks, favoured the defence and was far from ideal tank country, the Allied strategy was for the British and Canadian armies to maintain constant pressure and thereby draw in the bulk of the German armour while the Americans prepared to break out into the open country to the south and east. To achieve this a series of major operations – Epsom, Jupiter, Charnwood, Goodwood and Bluecoat – were mounted throughout June and July, wearing down the German strength in heavy attritional fighting. The American breakout, codenamed 'Cobra', commenced on 25 July and proved to be unstoppable; concurrently, at the northern end of the front, the Canadian First Army was pushing slowly but steadily southwards from Caen towards Falaise. With both their flanks now hanging in the air, the German armies were compressed within a pocket from which comparatively few escaped. By 21 August the fighting in Normandy was effectively over.

While these operations were in progress 79th Armoured Division received a new weapon which had not been employed during the D Day landings. This was the fearsome Crocodile, consisting of Wasp flamethrowing equipment fitted to a Churchill VII which was coupled to a two-wheeled armoured trailer holding 400 gallons of inflammable liquid. From the coupling, a pipe led under the tank's belly to emerge into the driving compartment where it joined the flame gun, which replaced the hull machine-gun. On leaving the flame gun the liquid was ignited electrically and propelled in a jet to a range of 120 yards, clinging to everything it touched. The propellant gas was pressurised nitrogen, housed in cylinders inside the trailer, permitting flaming for a total of 100 seconds in short bursts. Once the flaming liquid had been expended the trailer could be dropped, leaving its parent vehicle free to fight as a gun tank. Naturally, the Crocodile was hated and feared by the enemy, whose anti-tank gunners would attempt to knock out the trailer before it could be brought within range. Often, however, the appearance of a Crocodile was in itself sufficient to induce surrender; on the other hand, incidents occurred in which captured Crocodile crews were shown no mercy. The division possessed three Crocodile regiments – 141 Regiment RAC, 1st Fife and Forfar Yeomanry and 7 RTR – which together formed 31st Tank Brigade.

Another class of vehicle was added to the divisional armoury as a direct result of the fighting in Normandy, thanks to Lieutenant General Guy Simonds, the 41 year old commander of the Canadian II Corps, whose innovative ideas often complemented those of Hobart. The Allies had sometimes resorted to concentrated carpet bombing to blast their way through a sector of the enemy's front. This certainly obliterated anything in its path, but the huge crater fields made life very difficult for armoured vehicles trying to penetrate the gap, as any World War I tank crewman might have predicted. In planning Operation 'Totalize', the Canadian drive in the direction of Falaise on 8 August,

Simonds hit on the novel ideas of using bomb carpets to protect the flanks of the advance, and mounted his infantry in makeshift armoured personnel carriers. At this period the Allied armoured divisions' organic infantry were equipped with the M3 armoured half-track, but the infantry divisions had no APCs at all, save for their small tracked weapons carriers. Simonds therefore had the guns stripped out of a number of M7 Priest Howitzer Motor Carriages, which were then capable of carrying twelve infantrymen apiece; inevitably, these weaponless versions of the M7 were known as Unfrocked Priests. The idea worked so well that it was decided to convert turretless Sherman and Canadian Ram tanks to the role; with equal inevitability, the APC family so created were referred to a Kangaroos. Two APC regiments – 49 RTR and 1st Canadian APC Regiment – were formed with 150 Kangaroos each and attached to 31st Tank Brigade.

After Normandy, the first major operation involving 79th Armoured Division was the capture of the Channel Ports. Hitler had given specific instructions to the garrison commanders that they were to be denied to the Allies at all costs and each was protected by a cordon of fortified areas which included an anti-tank ditch, minefields and numerous concrete strongpoints. These were studied in detail and assault teams formed to deal with them – Crabs to clear lanes through the minefields and provide direct gunfire support, AVREs with SBG bridges and fascines to fill in the anti-tank ditches, and AVREs and Crocodiles to tackle the strongpoints. On their own, neither AVREs nor Crocodiles could guarantee success against the thick concrete structures; the AVRE's petard bombs might crack the concrete but they would not touch those inside; and, on the approach of a Crocodile, the defenders could retire into the inner chamber until it had finished flaming, then return to their fire slits. Together, however, they proved to be a deadly combination; once the AVRE had cracked open the structure the Crocodile would flame it and the burning liquid would flow inside, consuming the oxygen and forcing the defenders into the open.

Le Havre was assaulted by the 49th (West Riding) Division plus the 22nd Dragoons, A Squadron 141 Regiment RAC, 222 and 617 Assault Squadrons RE; and the 51st (Highland) Division plus B and C Squadron 1st Lothians and Border Yeomanry, C Squadron 141 Regiment RAC, 16 and 284 Assault Squadrons RE. The attack commenced at 1745 on 10 September and during the next two days the assault teams fought their way steadily through the defences, leaving the infantry to accept the garrison's final surrenders on 13 September. Faced with tough resistance and extensive minefields, the teams enjoyed mixed fortunes but, as on D Day, a sufficient margin had been allowed to ensure success. Personal initiative, too, played a major part; at Harfleur, for example, the AVREs of 222 Assault Squadron RE filled in an anti-tank ditch by felling several trees with their petards. One Scottish company commander, having taken over a captured bunker as his headquarters, answered a ringing telephone and found himself talking to a German, who he invited to surrender. The suggestion was indignantly refuted, but others sharing the line had no reservations and promptly emerged from a number of neighbouring strongpoints. They had felt secure behind their minefields and were dismayed when these were breached by the Crabs, which came as a complete surprise to them; understandably, it was the Crocodiles which had shocked them most, and they described their use as 'unfair' and 'un-British.'

The Americans had already taken St Malo and begun to assault Brest on 25 August. The garrison of this major port, consisting of 35,000 men under an extremely tough paratroop commander, Major General Hermann Ramcke, resisted so stubbornly that the attackers made little progress. At length General Bradley requested the assistance of Crocodiles to subdue the defences and in response B Squadron 141 Regiment RAC, commanded by Major I. N. Ryle, was attached to Major General W. E. Sands' US 29th Infantry Division for the reduction of Fort Montbarey. This consisted of an old casemated masonry fort within a moat, surrounded in turn by concentric lines of defence incorporating 40mm and 20mm gun positions and a minefield which included buried 300 pound naval shells. On 14 September, after American engineers had gapped the minefield, a two-tank Crocodile troop, covered by the squadron's gun tanks, led the infantry attack, burning up weapon pits as it went. One Crocodile blew up when it ran over a shell but the second continued until the American infantry had consolidated their gains around the fort itself. At this point the Germans began to emerge with white flags and the Crocodile, having exhausted its flame fuel and fired off all its 75mm ammunition, began to turn for home. As it did so it slid into an anti-tank ditch. A troop of Churchill gun tanks came forward to assist, but the first of these slid into another tank trap, the second shed a track and the third bellied in a crater. Observing this unexpected series of accidents, the enemy changed their minds, retired within the defences and opened fire again. Although disappointing in its eventual outcome, the day's fighting yielded 122 prisoners, plus two 50mm guns, one 105mm gun and two major strongpoints captured. All the ditched tanks were recovered despite sniper fire.

Two days later the battle was resumed: 'One troop of Crocodiles (Sergeant Decent), supported by direct fire from every available tank and self-propelled gun, crept up to the fort and rolled their flame over the moat. A gun tank pounded the main gate and three prisoners emerged. One was sent back to call for surrender – this was refused, so two more troops (Lieutenants C. Shone and T. P. Conway) gave the fort all the flame and HE they had, to the accompaniment of all guns at hand. Phosphorous and mortar bombs rained down and a 105mm gun pumped 200 more rounds at the gate. As the fire shifted to the northern edge, the sappers, under smoke, blew charges against the wall and the infantry went in. An officer with a white flag greeted them; he and his 30 men were being suffocated by smoke and phosphorous. The outhouses were blazing and after a little hand-to-hand fighting all was over. The Germans expressed their respect for flame and showed how effectively casemates had been penetrated and crews burned alive.' The fall of Fort Montbarey made Brest untenable and on 18 September Ramcke capitulated.

Meanwhile, Boulogne was attacked on 17 September by Major General D. C. Spry's Canadian 3rd Division, with assault teams provided by A and C Squadron 1st Lothians and Border Yeomanry, A and C Squadrons 141 Regiment RAC, 81 and 87 Assault Squadrons RE. The plan, conceived by Simonds, incorporated two phases. First, breaches were to be made in the outer crust of the defences by carpet bombing, through which infantry in Kangaroos would be carried forward to the point where cratering prevented further progress; they would then establish themselves while bulldozers carved routes across the devastated area. The second involved the assault teams, formed into three all-armoured columns, exploiting through the infantry and advancing into the town centre from different directions. The defences themselves were subjected to an additional heavy battering by the Second Tactical Air Force and RAF Bomber Command, while the Allied artillery support programme included participation by two fourteen- and two fifteen-inch coast defence guns firing across the Channel from England. In the event, the battle took the form Simonds had intended, although the severe cratering caused the assault teams as much trouble as the enemy. On the other hand, there were unexpected strokes of luck. One column, for example, was guided by a French taxi driver through the tangle of city streets, and the crews of two AVREs which had petarded the main gate of the Citadel into surrender, taking the German adjutant and 30 of his men prisoner, were considerably surprised when the next figures to emerge from the fortress were grinning Canadian infantrymen who had been ushered through a secret entrance by a member of the Resistance. The last of Boulogne's defences surrendered on 21 September and a further 9500 prisoners began their march into captivity; the Canadians sustained only 634 casualties.

Hardly had the fighting ended than 3rd Canadian Division began moving towards its next objectives, the coastal batteries between Cap Gris Nez and Calais, and the port of Calais itself. Here the assault teams were drawn from the same units that had stormed Boulogne, save that the AVREs were manned by 81 and 284 Assault Squadrons RE. Deliberate flooding of the surrounding area restricted the approach to a heavily fortified coastal corridor from the west but, against this, almost two thirds of the 7500-strong German garrison, consisting in the main of elderly or sick men, was required to man the coastal batteries. It was apparent as soon as the attack began on 25 September that they lacked the will to fight and, having witnessed the capabilities of the assault teams, most surrendered after a token resistance. By 1 October, at a cost of only 300 Canadian casualties, the entire area had been cleared. For the first time in four years, shipping in the Straits of Dover could operate without the menace of enemy gunfire. The Lothians, having captured the German flag flying over the Cap Gris Nez gun positions, presented it to the Mayor of Dover, which had itself been a regular target of the coastal batteries.

With the exception of Dunkirk, the garrison of which was to be contained and allowed to rot until the war ended, all the Channel ports were now in Allied hands. Yet, so thorough had been the German demolitions that it would be months rather than weeks before traffic would begin flowing through them and the Allied lines of communication stretched back some 400 miles from the German and Dutch borders to the Normandy beaches. Again, while the 11th Armoured Division had captured the Antwerp docks more or less intact on 5 September, these could not be brought into use until the enemy had been cleared from both banks of the Scheldt. Therefore, even while fighting for the Channel ports was in progress, plans were being made for opening the sea approaches to Antwerp. These included augmenting 79th Armoured Division's amphibious capability by the issue of the Buffalo LVT (Landing Vehicle Tracked, originally developed for the Pacific theatre of war – see Chapter 13) and its smaller cousin the Weasel. In British service the Buffalo could carry either 24 infantrymen, or one 17pdr anti-tank gun, or one 25pdr gun-howitzer, or one tracked Universal Carrier, or ammunition and supplies. 5 ARRE had, in fact already converted to the amphibious role and was joined by 11 RTR, both regiments being equipped with 100 Buffaloes.

The first area to be tackled was an enemy pocket on the south bank of the Scheldt, centred on the town of Breskens. When this was attacked on 6 October by the Canadian 7th Brigade little progress was made, partly because flooding had again been used to restrict the frontage, and partly because the defence was being conducted by the German 64th Division which, recruited largely from men on leave from the Eastern Front, was a very different proposition from the fortress troops encountered the previous month. However, during the early hours of 8 October 5 ARRE's Buffaloes ferried the Canadian 9th Brigade across the mouth of the Braakman Inlet (referred to as the Savojaards Plaat in the divisional history), covering the eastern flank of the pocket, and land-ed in the enemy's rear, achieving complete surprise. Major General Eberding, command-ing the German troops within the pocket, reacted quickly to the threat but was unable to prevent the Canadian 8th Brigade being similarly lifted into the beachhead two days later. Now under simultaneous pressure from north and south, he shortened his perime-ter, but one by one the towns in his possession fell to the Canadians and their support-ing teams of Crabs, AVREs and Crocodiles. By 3 November the pocket had been cleared.

Across the Scheldt was the South Beveland peninsula, connected to the main-land by an isthmus along which the 2nd Canadian Division was slowly fighting its way forward. To accelerate the capture of the peninsula the 156th Brigade, from 52nd (Low-land) Division, was carried the nine miles across the river in the Buffaloes of 5 ARRE and 11 RTR, accompanied by eleven DDs manned by B Squadron Staffordshire Yeomanry, in the pre-dawn darkness of 25 October, direction being maintained with the assistance of bursts of Bofors tracer fired at timed intervals from the south bank. A beachhead was established and rapidly expanded without difficulty, although only four of the DDs were able to surmount the muddy dykes and accompany the troops inland. On 27 October 157th Brigade, also from 52nd Division, was ferried across and that night the 2nd Cana-dian Division broke through the isthmus defences to join the Scots in overrunning the rest of the peninsula.

West of South Beveland was the heavily fortified island of Walcheren, covering the entrance to the Scheldt. Diamond shaped and measuring some twelve miles by nine, most of the island lies below sea level and is protected by dykes and sand dunes up to 100 feet high. The Germans were well aware of its strategic importance and had built numer-ous concrete coastal batteries containing guns of up to 220mm calibre, both among the dunes and inland; beach defences included mines, posts, hedgehogs, Element C and wire. Many of these defences were neutralised when, during early October, the RAF mounted a series of raids which breached the dykes in four places, allowing the sea to flood in.

Two landings were planned, the first based on Breskens and directed at Flush-ing on the south coast of the island with No 4 Commando leading in landing craft, fol-lowed by 155th Brigade (52nd Division) in Buffaloes, while the second, based on Ostend, was directed at Westkapelle on the west coast with 4th Special Service Brigade coming ashore in Buffaloes launched from LCTs, plus a strong armoured assault team, which would also land from LCTs.

Both landings took place on 1 November. That at Flushing went entirely according to plan, with the commandos having the good fortune to come ashore in the one area that had not been mined, while the Buffaloes sustained comparatively small loss-es, despite the heavy volume of fire directed at them.

At Westkapelle, however, the story was very different. The assault was to be delivered at the breach in the dyke just south of the town, but the sea approaches to this were covered by three major coast defence batteries. To the north of the gap were Battery W.17 at Domberg, armed with four 220mm French guns and one 150mm gun in open concrete casemates, and Battery W.15 on the northern outskirts of Westkapelle, armed with four British 3.7in AA guns in concrete casemates and two British three-inch AA guns in open emplacements, all of which had been converted to the coast defence role; south of the gap, between Westkapelle and Zoutelande, was Battery W.13, armed with four 150mm guns in concrete casemates, two 75mm guns in casemates and three 20mm AA cannon.

Because of this the landing force would have direct support from the 15in guns of the battleship HMS *Warspite* and the monitors HMS *Erebus* and *Roberts*, which would open fire at 0815, joined twenty minutes later by the First Canadian Army's medium and heavy artillery, firing across the Scheldt from Breskens. Escorting the landing craft during their run in towards the beach would be a Close Support Group containing 27 LCGs (Landing Craft Gun) and LCT(R)s (Landing Craft Tank (Rocket)).

Sea conditions were good but mist blanketing airfields in England had kept the promised air support grounded, although it was expected to clear. It also grounded the Bombardment Group's spotter aircraft so that *Warspite* and the monitors were forced to fire by dead reckoning. As a result, their heavy shells did less damage than was expected until the Artillery Air OP light aircraft arrived from Breskens; for a while, too, *Erebus* was forced to cease firing because of mechanical problems in her turret. Bursting shells began to fountain around the Support Group while it was still some four miles short of the shore, but seeing that the German fire had not been appreciably reduced, the gun and rocket craft closed in to conduct a deadly and one-sided duel with the enemy batteries. Their self-sacrifice succeeded in concentrating most but by no means all of coastal gunners' fire on themselves, and it eliminated a number of smaller posts and pillboxes, albeit at terrible cost; nine of the craft were sunk, eleven were put out of action with serious damage, and of the 1000 Royal Naval and Royal Marine personnel who provided their crews 192 were killed and 126 seriously wounded. At the critical moment, flight after flight of rocket-firing Typhoons roared in to strafe the defences, enabling the LCIs and LCTs to ground and disgorge their Buffaloes, which secured the shoulders of the breached dyke or passed through to assault Westkapelle from the rear.

The armoured assault team, carried in four LCTs, had the worst landing in the division's experience. 'One craft was hit repeatedly, the SBG bridge shot away from its AVRE and a fascine AVRE set on fire. By the efforts of its crew, explosives were jettisoned and the fire put out. The craft was later ordered to withdraw back to Ostend. The second LCT, *Cherry*, was hit hard astern and had many casualties, forcing it to withdraw and come in later. The third, *Bramble*, was able to touch down among broken boulders. The first AVRE to get ashore bellied (in the mud), so the craft pulled out and beached on sand further south. Two Crabs made heavy going of it but got up the beach; a bulldozer followed but the SBG was hit by a shell and the AVRE stuck. *Cherry* beached further south and offloaded all but one Crab, which was inextricably tangled up in the vessel's bridge which had been wrecked by fire. The bulldozer, in an attempt to recover the bellied AVRE off *Bramble*, sank itself in a quicksand.

'About this time Sergeant A. Ferguson (1st Lothians) fired eleven rounds of 75mm at the church tower east of Westkapelle which was being used as an observation post. The tower burst into flames and Germans came running out. Westkapelle had almost been cleared by the commandos but heavy shelling and mortaring of the beaches went on. All over the shore Buffaloes and Weasels were searching for an exit. Two Buffaloes loaded with ammunition were burning fiercely – in fact the beach was covered with smashed and burning vehicles and casualties were mounting.

'The fourth LCT, *Apple*, landed on the left. The first Crab got ashore but stuck when it tried to tow the second off. The bridge AVRE disembarked, laid its SBG over some bad going, crossed it, then bogged, which blocked the way of the second AVRE already on the bridge. In spite of all efforts both tanks were drowned by the rising tide.

'The tanks had found a small gap through the rocks but the sand was soft and they spoiled it completely for other vehicles. However, by dint of great perseverance and sterling work under shell and mortar fire, two Shermans, two AVREs, two Crabs and a bulldozer reached the village. They worked right through it dealing with roadblocks and houses and filling in craters with the bulldozed rubble and trees. That night the tide through the breach rose so high that the two Crabs were drowned.'

During the afternoon Battery W.13 surrendered to 48 Royal Marine Commando; one of its 150mm casemates had been cracked, killing the crew, and the remaining big guns had expended their ammunition. 41 Royal Marine Commando stormed Westkapelle village and Battery W.15 at about noon, then moved north to tackle Battery W.17 at Domburg, which gave up without a fight shortly after dusk. Resistance in this area, centred on Domburg village and a number of smaller concrete strongpoints among the dunes to the north, now became stiffer, and for the next six days the armoured assault team, reduced to two Sherman gun tanks and two AVREs, was fully engaged in supporting 41 Commando and No 10 Inter-Allied Commando as they fought their way along the coast towards Veere. Both AVREs eventually fell victim to mines, but the Shermans soldiered on to the end, firing no less than 1400 rounds of 75mm AP and HE. Later, Hobart received a personal note of thanks from Brigadier B. N. Leicester, commanding 4 Special Service Brigade: 'I want to let you know how very well your chaps did....The few tanks we got ashore were worth their weight in gold....In the north we had no close-supporting fire other than three machine-guns which were of little use against concrete. There the tanks consistently and successfully supported troop attacks on concrete by 75mm, the accuracy of which was a pleasure to watch – I saw them! Their Brownings too had a most heartening effect in keeping the enemy infantry to ground.'

After the coastal defences had been stormed and Flushing was captured, the remaining German resistance was centred on the fortified town of Middleburg in the centre of the island. As this was surrounded by flooded terrain, the garrison felt reasonably secure, but on 6 November the Buffaloes of A Squadron 11 RTR, with 7th/9th Royal Scots aboard, set out from Flushing to prove them wrong. The direct route lay along the banks of the Flushing–Veere Canal, but this was mined and covered by anti-tank guns and machine-guns. Under the guidance of a Dutch civilian, a more circuitous route over inundated countryside to the west was adopted. Thus far the enemy, drawn mainly from the 70th Division (nicknamed the White Bread Division because most of its men had stomach ailments), had put up a remarkably tough fight, but so surprised and demoralised were they by the sudden arrival of the leading infantry company in eight Buffaloes that they offered no resistance. The German garrison commander, Lieutenant General Wilhelm Daser, indicated his willingness to surrender, but not to a junior officer, a difficulty which was resolved by supplying the infantry company commander with a badgeless raincoat and according him local and temporary promotion to lieutenant colonel.

Between 1 October and 8 November the First Canadian Army had sustained 12,800 casualties, half of them Canadian, from all causes. In return, over 41,000 prisoners had been taken and both banks of the Scheldt had been cleared. Minesweeping had begun even before Walcheren surrendered and the first cargoes reached Antwerp on 26 November.

The 79th Armoured Division continued to provide specialist armoured teams for the whole of 21st Army Group, plus the US First and Ninth Armies, and were engaged in a variety of operations spread across a wide area. On 16 December a major German counter-offensive was launched through the Ardennes with the object of recapturing Antwerp, but the only elements of the division to become involved in the subsequent fighting, collectively known as The Battle of the Bulge, were the two Kangaroo regiments and one troop of 11 RTR's Buffaloes.

The counter-offensive failed, although it did delay Allied plans for an advance to the Rhine by approximately six weeks. After the threat had passed, 79th Armoured Division commenced detailed planning for Operation 'Veritable', 21st Army Group's eastward drive through the Reichswald Forest into the Rhineland. For this, the initial allocation of specialist armour was as follows:

15th (Scottish) Division: One regiment of Crabs, two squadrons of AVREs, two squadrons of Crocodiles and both Kangaroo regiments;
51st (Highland) Division: One squadron each of Crabs, AVREs and Crocodiles;
53rd (Welsh) Division: Two squadrons of Crabs and one squadron each of AVREs and Crocodiles;
2nd and 3rd Canadian Divisions: One squadron each of Crabs and AVREs, four squadrons of Buffaloes.

Following the heaviest preparatory bombardment fired by British artillery in World War II, Veritable commenced on 8 February 1945. Conditions were atrocious, continuous rain having produced deep mud, while on the northern flank the Germans breached the dykes protecting the Rhine flood plain, which now lay under five feet of water moving at a speed of eight knots. The Reichswald itself consisted of close plantations and was considered by some German officers to be tank-proof on its own account; in addition, it also contained a northern extension of the Siegfried Line, and while the concrete structures for this had not been completed, there were bunkers, minefields and anti-tank obstacles. Holding the area was the 84th Division, which had twice been bled white since the Normandy landings and had recently been reinforced with more units consisting of men with medical conditions; this was not expected to put up much of a fight, but in immediate reserve were troops of the German First Parachute Army who could be relied upon to contest every foot of ground.

During the early days of the battle the appalling going again provided the armoured assault teams with more problems than the enemy. Wallowing slowly forward through the morass, the vehicles bellied and towed each other free in agonising slow motion, with chilling rain sheeting down on them the while; at one stage it was estimated that three-quarters of the tanks in the forest were bogged down. Yet, somehow, lanes were cleared of mines by the Crabs, anti-tank ditches were bridged or filled in, and bunkers petarded or flamed into submission. On the right and in the centre only the

Churchill gun tanks of the infantry-support tank brigades and the Churchill-based Croc-odiles and burden-free AVREs were simultaneously able to cope with the mud and splin-ter their way through the trees. The commander of one captured feature went so far as to pass the outraged comment: 'We had never thought that anyone in their right mind would use tanks in this forest; it is most unfair!' On the left, where the flooding was total, the Canadians mounted their attacks in assault boats, supported by fire from Buffaloes, and were supplied by Weasels and DUKWs.

It was this ability to retain mobility that the Germans found most unsettling. As the rains became less frequent and the floods started to subside, they rushed nine more divisions to the threatened area, drawn from the American sector of the front. This suit-ed the Allies very well for on 23 February the US Ninth Army, forming the right wing of 21st Army Group, seized crossings over the river Roer at a cost of less than 100 casual-ties and, breaking out of its bridgeheads a week later, its armour reached the Rhine on 2 March. The effect was to isolate First Parachute Army, still locked in its bitter struggle with the British and Canadian divisions to the north, so that von Rundstedt, the Com-mander in Chief West, had no alternative other than to withdraw what remnants he could across the river. Together, the Reichswald battle and the American offensive had cost the German Army some 90,000 men. The British sustained 10,330 casualties, the Canadians 5304 and the Americans 7300.

Plans were already in hand for the crossing of the great waterway. These includ-ed use by 79th Armoured Division of a device cloaked in such secrecy that, although units had been equipped with it since 1942, it had never been used operationally by British troops, an omission described by Fuller as the greatest blunder of the war. The device was a British invention known as the Canal Defence Light (CDL) and it consist-ed of an M3 Lee/Grant chassis and hull fitted with a specially designed turret housing a 13 million candlepower carbon arc the intense light from which was reflected through a narrow slit controlled by the rapid movements of a mechanically driven shutter. The flickering effect of the light induced temporary partial blindness, sometimes accompa-nied by nausea, loss of balance and disorientation, and it also prevented the enemy iden-tifying its source or even whether it was moving or stationary. Tactically, it was possible for troops to advance towards their brilliantly illuminated objective, yet remain com-pletely invisible in the dense black space between two CDL beams. The Americans were also interested in the idea and had formed CDL units, although they codenamed the vehicle the Shop Tractor. Once again, however, obsessive secrecy had prevented its use until, even as 21st Army Group was completing its preparations for crossing the Rhine, the US 9th Armored Division seized the Ludendorff Bridge at Remagen by coup de main and a CDL unit was successfully employed to defend the structure against attacks by Ger-man frogmen; even this local usage had required the personal sanction of General Dwight D. Eisenhower, the Allied Supreme Commander. On the British sector the CDLs were to be used in a similar fashion, as well as providing directional light for the Buffalo cross-ings, requiring B Squadron 49 RTR, one of the original CDL units, to abandon its Kan-garoos and retrain in the role.

For the Rhine Crossing, codenamed 'Plunder', and the subsequent expansion of the bridgeheads secured, 79th Armoured Division effected the greatest concentration of specialist armour since D Day. This included two DD regiments (Staffordshire Yeomanry and 44 RTR) and three additional Buffalo regiments (1st Northamptonshire Yeomanry, 1st East Riding Yeomanry and 4 RTR). The lessons of South Beveland and Walcheren had been learned and, to assist the DDs in climbing the floodbanks, they would be accompanied by a number of specially adapted Buffaloes that would unroll a chespale carpet ahead of them where necessary. As few permanent defences existed on the east bank of the river there was no immediate requirement for AVREs and assault regiments RE were given the task of manning motorised rafts which would be used to ferry non-DD tanks and other vehicles across in the wake of infantry divisions.

The allocation of units from 79th Armoured Division to the higher formations involved in the crossing was:

BRITISH SECOND ARMY

British XII Corps: two regiments plus one squadron of Buffaloes, three assault squadrons RE manning motorised rafts, one regiment of Crocodiles, two squadrons of Kangaroos, one Crab regiment, one half-squadron of CDLs and one DD regiment.

British XXX Corps: two regiments of Buffaloes, three assault squadrons RE manning motorised rafts, three squadrons of Crocodiles plus one in corps reserve, one squadron of Kangaroos, one squadron of Crabs plus two in corps reserve, one half-squadron of CDLs and one DD regiment.

US NINTH ARMY

US XVI Corps: one squadron of Crocodiles, one squadron of Crabs.

Plunder was the sort of huge set piece operation in which Montgomery excelled. The river crossing, supported by air attacks and the fire of 3300 guns on a 25-mile frontage, commenced at 2100 on 23 March. The six understrength German divisions on the east bank could only offer weak resistance and by morning bridgeheads had been established with very few casualties; before the enemy could recover his balance the US XIX Parachute Corps (British 6th and US 17th Airborne Divisions) had been dropped behind him. Armour began to flow across the river and, day by day, the bridgeheads were expanded until the point was reached when the German Army, with nothing in reserve, was unable to contain them. The war had only weeks to run.

For the Royal Tank Regiment the Rhine crossing was memorable in a number of ways. Lieutenant Colonel Alan Jolly, commanding 4 RTR, crossed with the same brown, red and green flag flying from his Buffalo that the 17th Battalion Tank Corps had flown when their armoured cars reached Cologne at the end of World War I. On 26 March 11 RTR's Buffalo crews on the east bank were rounded up to meet one of their vehicles that was carrying more brass hats than they had ever seen in one place. The men were deadly tired and hollow-eyed, their battledress crumpled and their boots plastered with mud. Their days involved many hours of continuous work, with a few precious intervals for rest and sleep, and the arrival of the generals was less than welcome. As the VIPs clambered down, however, they began to take more interest. Major General Hobart, their divisional commander, was of course a familiar figure; equally familiar, although not many had seen him in the flesh, was the Army Group commander, Field Marshal Sir Bernard Montgomery, casually dressed and wearing their own black beret; some probably recognised Lieutenant General Neil Ritchie, commanding XII Corps, but not many could have put a name to Field Marshal Sir Alan Brooke, Chief of the Imperial General Staff, or to General Sir Miles Dempsey, commander of the British Second Army. Yet it was a cherubic figure wearing only a lieutenant colonel's insignia on the shoulder straps of his British Warm, and the badge of 4th Hussars, his old regiment, in his standard service dress cap, who attracted most attention. As he called the men to break ranks and gather round delighted grins began to spread and there were astonished exclamations of: 'It's Winnie!' The Prime Minister congratulated the men on their efforts and Montgomery, completely relaxed after the success of the operation, urged them to take advantage of the situation: 'Go on, ask him for a cigar!' At that precise moment, it would have been strange for one so steeped in a sense of history if Churchill had not reflected on how armoured warfare had developed since he had encouraged its first painful beginnings some thirty years earlier. Later, he visited the US Ninth Army's bridgehead, causing near heart failure among senior American officers by repeatedly exposing himself to danger.

The CDLs of B Squadron 49 RTR had also fulfilled their promise. They naturally attracted much of the enemy's fire, and consequently were unpopular neighbours, but they were difficult to locate and only one tank was lost. From 25 March until 6 April their role became the night defence of the pontoon bridges which the engineers had worked at top speed to complete. Three frogmen were exposed by the flickering beams and captured, and a large number of floating objects were engaged and sunk. Some of the latter exploded and while most of these devices were primitive mines consisting of logs to which charges had been strapped, one was thought to be a midget submarine, of which the Germans had several types. The CDL squadron followed up the advance into central Germany and was in action again during the crossing of the Elbe at the end of April.

During the last month of the war in Europe 79th Armoured Division was again dispersed across a very wide area, assisting in the liberation of those parts of Holland still under German control, driving north to participate in the capture of Bremen and Hamburg, and east until contact was established with the Russians. One of the most remarkable facts about the division's history was that the nature of both its equipment and the operations it undertook was successfully concealed from the general public until after the Rhine Crossings. The media then received appropriate releases, resulting in a series of sincere if over-enthusiastic tributes, some of which fell just short of claiming magical powers.

At this period the divisional strength amounted to 21,430 men and 1566 armoured fighting vehicles; the strength of a conventional armoured division was about 14,400 men and 350 AFVs. 79th Armoured Division was a unique formation, although a similarly equipped but much smaller assault brigade supported the British Eighth Army's final offensive in Italy. 'Hobo's Funnies' were disbanded in 1945, and although assault engineering techniques have reached new levels of sophistication, no formation of the same size and scope as Hobart's division has been formed since, for the good reason that the circumstances that called it into being have mercifully never been recreated.

'Hideous Nights Pressed on Days of Horror'
Fighting in the Gothic Line, Italy, Autumn 1944

All tactics are influenced by terrain, and that of Italy unequivocally favoured the defence. From the Apennine chain, which forms the country's spine, rivers run east and west into, respectively, the Adriatic Sea on the one hand and the Tyrrhenian and Ligurian Seas on the other, and these are separated by ridges running inland from the narrow coastal strip, increasing steadily in height until they reach the main body of the mountain range, so that armies attempting to advance along a south-north axis, as did those of the Allies between 1943 and 1945, do so against the grain of the landscape.

Furthermore, Italy's turbulent history had ensured that many of the larger villages and country towns were built on hilltops and walled for their own defence. Not only were these difficult to capture, they frequently dominated the only roads and provided ideal artillery observation posts. Again, where the step-terrace method of agriculture was employed, this inhibited cross-country movement, and the vineyards, with their narrow fields of vision between the lines of vines, gave ideal cover for tank hunting parties. The mountain roads, narrow and few in number, were natural targets for the enemy rearguard's demolition teams, who could blow bridges, culverts and embankments at will, slowing the pace of their advancing opponents to a crawl. Of one stage of the war in Italy it has been written that the best weapon in the Allies' advance guard was not the armoured car or the tank but the bulldozer.

Furthermore, the role of the tank had undergone a subtle but profound change since the beginning of 1943. This arose following the introduction of close range, hand-held, hollow charge infantry anti-tank weapons - the bazooka for the Americans, the PIAT for the British and the Panzerfaust for the Germans. In close or broken country these enabled a man to lie in cover until a tank was within range, then blast his bomb into its flank or rear. The infantryman therefore lost his former awe of the tank and for the second half of World War II the latter, once Queen of the Battlefield, performed best as part of an all-arms team. In practice, this meant that in scrub, standing crops or built-up areas it was the infantry who led the advance, flushing out the tank hunters as they went, although in more open country the tanks took the lead as usual.

Restricted as was the employment of tanks in Italy, one has only to compare the terrible loss of life in the tankless mountain battles around Monte Cassino with casualties sustained elsewhere in the campaign to appreciate the extent that they were able to influence the course of the fighting; there were, too, times when the presence of tanks was critical to the survival of the hard pressed Salerno and Anzio beachheads. Even when conditions were completely static, as they were for many months, the Allies' Shermans and M10 tank destroyers made a valuable contribution in the role of supplementary artillery. It was, however, in the infantry support role that the tanks were mainly employed throughout the campaign, every attempt to exploit a potential breakthrough with armoured divisions being thwarted by the combination of terrain, demolitions and stubborn defence until the last days of the war.

There are few more graphic illustrations of the all-arms assault technique in action than the Allies' prolonged struggle to break through the Gothic Line in the autumn of 1944. This defensive zone, up to ten miles deep in places, ran from the Magra valley, south of La Spezia on the west coast, through the Apuane Mountains to a chain of strongpoints guarding the Apennine passes, thence to the Foglia valley and a point on the Adriatic between Pesaro and Cattolica. It contained every obstacle that human ingenuity, backed by steel and concrete, could devise, including bunkers, dug-in Panther tank turrets on ground mountings, minefields, wire entanglements, cleared and registered artillery killing grounds and mutually supporting anti-tank and machine-gun posts; while as mobile reserve there were the tanks and assault guns of 26th Panzer Division, two Tiger battalions and several self-propelled Panzerjäger battalions.

On the eastern sector the British Eighth Army the infantry attacks had the direct support of 21st Tank Brigade (12 RTR, 48 RTR and 145 Regiment RAC) and 25th Tank Brigade (North Irish Horse, 51 RTR and 142 Regiment RAC). Within regiments, each of the three fighting squadrons consisted of two troops and an SHQ troop equipped with Churchills, plus two troops equipped with Shermans. During an attack, the more heavily armoured Churchills formed part of the first wave while the lighter Shermans followed and provided direct gunfire support. Whenever possible, the tanks dealt with the enemy's machine-guns while the infantry neutralised his anti-tank guns. More often than not, the artillery's Forward Observation Officer (FOO) travelled with the tanks, his guns on immediate call to deal with targets as required. If the tank squadron leader was lucky, one or more troops of M10 tank destroyers would be placed under his command, and these would usually be positioned on the best point of vantage from which they could either protect the tanks from the enemy's armour or eliminate his observation posts in church towers or other high buildings. Provided they were available, further elements could also be introduced to the team, including bridgelayers and AVREs. Extensive use was also made of the Ark, which was simply a Churchill hull, fitted with folding ramps at either end, that could be driven bodily into an anti-tank ditch, crater or even the bed of a mountain stream, thus creating a causeway across the obstacle; if the obstacle was unusually deep, a second Ark could be driven on top of the first. In this way, every component arm of the team had the support of at least one other arm. Overall command of the operation rested with the senior infantry officer, unless the enemy counter-attacked with armour, when the tanks would fight as a conventional armoured squadron. Once an objective had been secured the tanks would remain with or near the infantry as a safeguard against counter-attack, and perhaps carry out a forward rally during which fuel and ammunition would be replenished, until the infantry's own anti-tank guns had been brought forward and emplaced. Once the position had been consolidated, the tanks would be released by the infantry.

The battle began on 28 August and raged on throughout September, lasting twice as long as the Second Battle of Alamein. It was bitterly contested throughout, and one tank driver of 51 RTR recalled the prolonged ordeal as being one of 'Hideous nights pressed on days of horror – we lost men, we lost tanks, almost we lost hope of survival.' The three regiments of 21st Tank Brigade lost a total of 62 crewmen killed, 227 wounded and seventeen missing. The brigade's tank losses amounted to 52 Churchills, 29 Shermans and four Stuarts from the regimental reconnaissance troops; of these fifteen

Churchills and 21 Shermans were repairable, producing a net loss of 37 Churchills, eight Shermans and four Stuarts. The greatest number were knocked out by the enemy's anti-tank guns (22), followed by tanks (21), assault guns and Panzerjäger (15), artillery fire (14), Panzerfausts (8), mortar fire (3) and mines (2). Throughout the battle the tanks expended 2828 AP rounds, 9632 HE shells, 2462 smoke shells and almost one million rounds of machine-gun ammunition. Typical of the tributes paid to the tank crews was that contained in 128th Infantry Brigade's operational summary of the battle: 'Day after day their Churchills forced positions and supported our infantry over appalling tank country. Undaunted, squadron leaders on foot led their tanks up seemingly impossible slopes. One tank actually slipped over and crashed two hundred feet down into a ravine, having turned over six times in its descent.'

All of the many actions fought involved solving numerous local tactical problems, and that in which B Squadron 48 RTR supported the Canadian 48th Highlanders on 4–5 September was typical. Earlier actions had left the squadron a little thin on the ground, so that it consisted of six Churchills and one Sherman only, organised in two troops with a small squadron headquarters. In support, however, were nine M10s, two troops of AVREs and the artillery.

During the afternoon of 4 September the infantry ran into machine-gun fire raking a reverse slope and requested assistance. The squadron leader sent forward one troop to assist. The troop leader was engaging one machine-gun when his tank was hit on the idler wheel by an armour piercing round. The enemy weapon fired twice more, missing, and the Churchill replied with its six-pounder. The troop leader had already ordered his driver to move forward into the cover of a re-entrant when the tank was hit again and this time began to burn, although all the crew escaped unhurt. It seemed as though the way forward was covered by one or more assault guns or Panzerjäger (referred to collectively as SPs, that is self-propelled guns, in Allied accounts) and a fresh plan was agreed by the squadron leader, the commander of the Highlanders' D Company, the FOO and the commander of the M10 battery.

'The machine-guns were to be knocked out first, then the infantry were to stalk the SP guns and destroy them with PIATs. The clearing of the machine-guns was to be left to the tanks. The M10s were manoeuvred carefully into fire positions to take on the SP guns. The SP on the right was observed to be a Panther and the range 2000 yards. It was engaged and moved off at once, first behind a house and later into some trees. The M10s fired about a dozen rounds and the artillery fired to encourage him to move into view again. What success was achieved is hard to say, but later two brewed-up Panthers were examined in the area. The gun or tank on the left could not be observed so we put down an artillery concentration on the area.'

Light was now failing but it was decided to renew the attack under cover of darkness, with the infantry leading and the tanks joining them on the objective at first light. Just after midnight a disquieting message was received from the infantry brigade headquarters. Interrogation of a prisoner taken on an adjacent sector revealed that the enemy had already anticipated this course of action and had deployed sixteen Panzerfaust teams across the line of advance with orders that they were to let the Canadian infantry pass and then open fire on the tanks at close quarters when they appeared. This placed the squadron leader on the horns of a dilemma. Either the tanks remained in comparative safety where they were, in which case the Highlanders would be left without their support at first light, or he could push on and risk losing what remained of the squadron in a blind close-quarter fight. After discussing the situation with the infantry battalion commander on the radio he opted for the latter, urging caution on the troop leaders and briefing the accompanying infantry on the degree of protection expected from them. The tanks moved off at 0130 and by 0415 had rejoined the Highlanders and snugged down in fire positions among the companies, without encountering the slightest opposition.

'At 0530 D Company continued their advance supported by tanks and arrived at Point 70, with no trouble except sniping and mortaring, by 0930. D Company pushed on down the hill to the creek running north and south but got no further owing to Spandau fire from carefully sited positions along the Coriano – Riccio Marina road. The squadron leader lost touch with them through his wireless set but got in touch through C Company's, who were astride the ridge at Point 70. C Company commander was able to locate some machine-gun positions and the squadron Second in Command and the FOO spotted a 75mm [wheeled] gun, a Tiger camouflaged as a haystack and a number of other suspicious objects. The squadron leader called an orders group at the church on Point 70 and allocated sniping targets to the six-pounder [Churchills], the 95mm [Close Support Churchills] and M10s and a linear shoot for the artillery along the road. This was quite successful. Nine men ran away from the 75mm and the barrel cocked up in the air. The Tiger was hit and later brewed up. A lorry drove away from a house that was engaged, and the Spandaus on the right stopped firing. Lieutenant Greenwood's tank, whilst engaging a target, was hit and both tracks were broken and the hull damaged. All attempts to tow this tank out of action were met with fire from the enemy, and when the two-inch smoke mortar was used to cover operations such a bad stonk came down that it was left there. On the night of 5/6 September a small German booby trap was placed under the tank, presumably by a patrol. This sector was taken over by 5th Canadian Armoured Division, and as B Squadron had only three battleworthy tanks left they were placed in reserve and withdrew about 300 yards. By 1500 eleven tanks were battleworthy, but the regiment was then pulled out for rest and reorganisation.'

Between 18 and 20 September C Squadron 48 RTR provided tank support for the West Nova Scotia Regiment (WNSR) in the San Fortunato area, and their account of the action reveals the full extent of the danger presented by a well-emplaced Tiger with a good field of fire.

'Captain Hoad received orders for the attack from the officer commanding WNSR at his headquarters in the area of San Lorenzo at about 2200 and orders were given to troop commanders at about 2230. Information was received that the enemy were holding the high ground Point 152 – San Fortunato – Covignano in strength, and that his forward elements were approximately on the line of the west bank of the river Ausa. Our own leading troops in this sector were in contact with the enemy and the Carleton and York Regiment (CYR) were dug in on the east bank of the river. Patrols had crossed the river and encountered heavy machine-gun fire from positions on the opposite bank.

'The intention was that the WNSR supported by C Squadron should attack and capture Point 152. At the same time A Squadron was to support the Hastings and Prince Edward Regiment onto San Fortunato, and the R 22e Regiment [a French Canadian unit] without tank support, were to take the wooded feature on the left, the Villa Francolini. This last attack was to be timed to take place a little later. The method was that 1 Troop [Churchills] and 2 Troop [Shermans] should move across the river immediately behind the WNSR, as soon as the CYR had established a bridgehead and the Royal Engineers had made two crossings. 1 Troop was then to support the right leading company, and 2 Troop the left leading company, up onto the final objective. This was easily recognisable as the centre of three high pimples commanding the river valley to the south-east. It was hoped that by first light (0550) the infantry and tanks would have reached their first objective, the line of the lateral road running along the east side of the San Fortunato feature. All possible support was to be given by 4 Troop (Shermans) and Squadron Headquarters, which included two Churchills fitted with 95mm howitzers, from the spur immediately north of San Lorenzo.

'It was not until about 0200, September 19, that the necessary bridgehead was established over the river Ausa and the tank crossing completed, when 1 and 2 Troops received orders to move. They soon caught up with the WNSR and continued across country. During this move along narrow tracks and in poor light one tank each of 1 and 2 Troops became ditched, but the reserve Churchill, which had been held ready for such an emergency, joined 1 Troop at once, though 2 Troop were compelled to go into battle with only two Shermans. Lieutenant R. E. Wife (1 Troop) made a reconnaissance on foot to find a crossing of the wadi in this area, but they had to move north to A Squadron's crossing 1500 yards downstream. This crossing was badly damaged but a fascine was laid across the wadi by a squadron of the Assault Regiment RE, representatives of which happened to arrive on the spot at this time.

'Shortly after first light the tanks joined the infantry just behind the first objective, which was to serve as the start line for the second phase of the operation, the attack on Point 152 itself. The troop commanders now contacted A and B Company commanders to tie up details of this second phase, in which 1 Troop on the right was to support A Company and 2 Troop on the left was to support B Company. It was agreed that, owing to the closeness of the country, one tank should support each platoon and that, as the tank would not have the support of the other tanks of the troop, it was essential that the infantry should make themselves responsible for close protection of the tank, especially against bazooka men.'

The area between the start line for the next phase of the advance and the final objective was subdivided by a first intermediate objective and a second intermediate objective, the latter being a straggling line of houses. One tank from each troop was to support one platoon from each company as far as the first intermediate objective, after which a second tank from each troop with a second platoon was to push on to the second intermediate objective.

'On the right Sergeant Paice (1 Troop) supported the leading platoon of A Company onto the first intermediate objective. His traverse gear became damaged when he was still on the start line, which was a sunken road with a steep bank about five feet high on the enemy side. The six-pounder gun struck a tree, the result of which was that the turret would only traverse to the left. From this position, however, he was able to cover the infantry, and 1 Troop commander moved over to the right flank to give further cover from there. The infantry reached the first intermediate objective. On the left Lieutenant Fox (2 Troop) went up to the first intermediate objective with the leading platoon of B Company. His forward Browning jammed, however, and immediately afterwards the feed lever of his coaxial Browning broke, a round jammed in the 75mm, and just then he was fired on at short range by a bazooka man, who missed badly. Lieutenant Fox charged him down and then came back to attempt to repair his guns.

'Over to the right Sergeant Clinton in the third tank of 1 Troop now supported the second platoon onto the second intermediate objective where he remained with them. He was assisted by 1 Troop commander whose six-pounder firing cable broke at this juncture, putting the gun temporarily out of action. He also had to change a Besa barrel owing to a split case. Lieutenant Wife now returned to company headquarters to receive orders for the final attack. On the left Sergeant Catchpole (2 Troop) was to support the second platoon of B Company onto the second intermediate objective. Smoke was laid and maintained for ten minutes, and Sergeant Catchpole kept up a withering fire at targets in this sector, but counter-fire was heavy and the infantry were unable to get on. At this point a message was received that A Company on the right had reached the second intermediate objective and Sergeant Catchpole was recalled to company headquarters.

'It was decided to put in the attack on the right supported by 1 Troop commander and 2 Troop sergeant. All available strength of both companies was mustered for this attack. Sergeant Catchpole, however, failed to cross the start line as his engine stalled just before the sunken road and could not be restarted. Lieutenant Wife moved with the infantry up to the second intermediate objective, where Sergeant Clinton was still in position. The latter now reported a Tiger tank firing from the summit of Point 152, and Lieutenant Wife moved up the slope to get better observation.'

The Tiger was firing from an almost unassailable position. It had squeezed itself between two buildings and was protected in front by a wall which was apparently five to ten yards in front of the tank. Below the wall the ground fell away very sharply and the only target which the Tiger presented was the gun barrel and the very top of the turret, which was just visible over the wall. Some suspicion existed that there were, in fact, two Tigers up on the ridge, although they were never seen to fire at the same time.

'Meanwhile, artillery support was requested to deal with the Tiger. Sergeant Paice gave Lieutenant Wife's tank cover as he moved forward and both tanks were fired on though not hit. Lieutenant Wife decided to leave Sergeant Clinton where he was with the infantry, as this was one of the very few positions, apparently, which was not open to direct fire from the Tiger. Sergeant Paice also managed to find a position which was not entirely exposed, but the enemy tank could not be located so that it was difficult to find a hull-down position.

'Lieutenant Fox had now cleared his gun and withdrew a short distance in order to try to spot the Tiger and bring fire to bear on it. For this purpose he dismounted from his tank. Meanwhile, Sergeant Catchpole had sent his co-driver to contact Lieutenant Fox's tank and get a tow. No sooner had the co-driver left the tank than it was hit and brewed up. The Tiger now brought down heavy HE fire in the area of Lieutenant Fox's tank, which temporarily prevented him from remounting, and shortly afterwards the tank was hit by AP fire and immobilised.

'Both the infantry, who were being mortared badly, and the tanks were now unable to move forward, and Lieutenant Wife contacted A Company commander and told him that he must have tank reinforcements. 4 Troop and Squadron HQ had been unable to give effective support from their positions owing to the poor visibility through dust and smoke, and in answer to Lieutenant Wife's call for further tank support 4 Troop was now sent forward to join 1 and 2 Troops.

'Lieutenant Wife had now moved over the road in the area of the second intermediate objective, but his tank was continually fired on by the enemy tank from the high ground above him, and whichever way he moved he was unable to avoid its fire. As the six-pounder was temporarily out of action and he could not locate the Tiger exactly, he ordered the crew to dismount until the situation became easier. The Tiger then opened up with direct HE fire on the crew without, however, causing any casualties. About this time Sergeant Clinton's tank suffered a direct hit on the turret with a mortar bomb and the commander was temporarily blinded.

'4 Troop (three Shermans), commanded by Lieutenant Buckley, arrived in the area about 1600, but could not find the way to the other two troops. Lieutenant Wife endeavoured to contact the tanks on foot and guide them to the right location. They eventually reached the area of A Company HQ, but lost one tank commander en route, who had been sniped and killed at short range. No sooner had they arrived in the area than the Tiger opened up on them, hit the troop corporal's tank and caused it to brew up. The crew escaped. Lieutenant Buckley's tank was hit next and disabled, and Buckley himself was wounded by HE fire after bailing out.

'By this time the two tanks of Squadron HQ had arrived on the scene and were promptly engaged by AP from the Tiger. The tanks endeavoured to find protected positions, but the enemy tank had an uninterrupted field of fire across the valley and the only chance was to seek protection behind a building. The six-pounder Churchill was twice hit and disabled, the left track being hit and broken by one shot which first went through the corner of a house. The 95mm Churchill suffered only slight damage to the bogies and was able to pull up in comparative safety behind a house which A Company commander was using as his headquarters. The manoeuvre could hardly be expected to arouse the gratitude of the infantry in the building, but in fact Captain Kingsford found the company commander quite agreeable and apparently unperturbed by the strong probability of the tank attracting fire on his headquarters.

'By now all the tanks were short of petrol and 1 and 2 Troops were very short of ammunition, and it was clear that no further attack could be made that evening with the forces already on the ground. Shortly after dark, orders were received to consolidate the positions already held and to pull back the tanks (for replenishment) at 0500 on 20 September.'

The day's fighting had cost the squadron one Churchill and four Shermans written off; incredibly, personnel casualties amounted only to one killed and two wounded. At last light C Squadron's Canadian FOO had been reasonably certain that his guns had finally neutralised the Tiger, but as the tanks pulled back in the half light of dawn they were pursued by several high velocity rounds indicating that this was clearly not the case. In the meantime, an attack in brigade strength, supported by 48 RTR's B Squadron and two squadrons from 145 Regiment RAC, had been put in further along the ridge during the night and succeeded in capturing most of the high ground.

At 0945 on 20 September, having completed its replenishment and now seven tanks strong, C Squadron moved forward again to rejoin the WNSR and supported one of the latter's companies in an attack on the Villa Francolini position. This time the Tiger was not in evidence and after a stiff fight the feature was taken, completing the capture of the ridge. From the summit Rimini was visible to the right, with the coastline stretching away in the direction of Venice, while ahead lay the more open plains of Lombardy.

The capture of the heavily fortified San Fortunato ridge finally cracked open the Gothic Line on the Eighth Army sector. However, expert demolitions and the stubborn German rearguards delayed any possible exploitation into the plains until the autumn rains had turned them into a quagmire and large-scale movement became impossible. Once again, the front became static, this time along the line of the river Senio.

During the winter the Allies prepared for what was to be the final offensive of the campaign. 25th Tank Brigade became the 25th Armoured Engineer Brigade, consisting of 51 RTR with two squadrons of Crocodiles and one with Crabs, and 1st Armoured Engineer Regiment RE with AVREs and bridging equipment. Further additions to the Eighth Army's order of battle were Kangaroo APCs, manned by squadrons from several armoured regiments, and LVTs. When the offensive began on 9 April 1945 no difficulty was experienced in breaking the Senio line; the Argenta Gap, the last major obstacle between Eighth Army and the Po valley, was forced by a combination of direct assault and an amphibious flanking attack across Lake Comacchio in LVTs. Beyond these points the German defence had little depth and even the slow Churchills participated in the subsequent pursuit to the point that their steel bogies ran hot.

The Armoured Car
in World War II

Whereas in World War I armoured cars had been employed almost exclusively as combat vehicles, in World War II they were primarily employed as the eyes of their armies, providing senior officers with a constant stream of reports on the enemy's movements, and in this context their radio sets were their most important weapon. In addition, their versatility enabled them to be employed in a variety of roles, including route and bridge reconnaissance, seize and hold operations to secure specific tactical objectives, controlling artillery fire, convoy and prisoner escort and, in the wide spaces of the North African desert or Russia, dominating no-man's-land.

It was in this last role that the 11th Hussars, commanded by Lieutenant Colonel John Combe, were first employed when Mussolini declared war on Great Britain and France on 10 June 1940. The 11th Hussars was the armoured car regiment of 7th Armoured Division and the declaration found it lying close to the frontier wire separating Egypt from Libya. All three of its squadrons, equipped with 1924 pattern Rolls Royce and Morris armoured cars, were immediately ordered to cross into hostile territory with instructions to raise hell, keep the enemy on edge and report what he was up to. These were tasks for which the squadron and troop leaders, delighted to be off the leash, were temperamentally well suited, and they set to with a will.

No difficulty was experienced in breaking through the wire, which parted under slow, sustained pressure from the cars' armoured radiator flaps. On the night of 11 June Lieutenant Miller's troop ambushed a convoy of four lorries heading for the small post of Sidi Omar. After a brief exchange of fire two Italian officers and 50 Libyan soldiers surrendered, the former protesting vehemently about the use of 'neutral' Egyptian territory as a base for aggressive action. The effect of the incident was immediate, for when another patrol probed towards the post two days later it was found abandoned and partly gutted.

On 13 June a two-car Hussar patrol also probed the defences of another frontier post, Fort Maddalena, to the south. The cars were driven off by heavy machine-gun fire then forced to take violent evasive action when they were subjected to an hour-long attack by six Caproni bombers and nine Fiat fighters. The following day the whole of A Squadron was detailed to attack the fort in the wake of an air strike by Blenheim fighter-bombers. After the bombs had fallen the few remaining occupants surrendered to the Hussars after a token resistance, most of the garrison having been withdrawn during the night.

Simultaneously, Fort Capuzzo, covering the main highway into Libya west of Sollum, was captured by the 1st King's Royal Rifle Corps, a motor battalion, supported by cruiser tanks belonging to 7th Hussars. Some miles to the north, however, the defenders of the fortified camp at Sidi Azeiz not only held their own against the 7th Hussars' light tanks and the Eleventh's armoured cars, but also launched a counter-attack on the latter with a squadron of L3 tankettes. These amounted to little more than tracked machine-gun carriers and were incapable of inflicting much serious damage on another

AFV; after one had been knocked out with a Boys anti-tank rifle the remainder broke off the engagement.

On 15 June one of the 11 Hussars' patrols ran down an enemy staff car containing an engineer general and two lady companions, evidently out for a spin in the desert. Despite the general's protestations of outrage, he and his friends were escorted across the frontier. He was also relieved of his briefcase, which contained a set of detailed plans of the fortifications surrounding Bardia; these would be put to excellent use when the fortress was stormed the following January.

Obviously, such events could hardly occur without the Italian High Command insisting that some sort of order should be restored along the border. On 16 June a detachment consisting of two lorried infantry companies, an L3 squadron and four guns was despatched into the danger zone with orders to 'Destroy enemy elements which have infiltrated across the frontier and give the British the impression of our decision, ability and will to resist.' The column immediately fell foul of 11th Hussars' prowling cars. In response to their contact report two cruiser and three light tanks of 7th Hussars, together with a troop of two-pounder anti-tank guns, converged on the scene. After two L3s were knocked out the column commander took the fatal decision to form square with the lorried infantry in the centre, the tankettes on the flanks and a gun at each corner. As a result, the so-called Battle of Ghirba degenerated into something resembling a massacre. The L3 squadron commander, a former cavalryman, bravely launched a counter-attack but one by one his little vehicles were knocked out. The British tanks and armoured cars then circled the square, raking it with their fire. As they were to do throughout the campaign, the Italian artillerymen continued to serve their guns until they were shot down. Now defenceless, the infantry broke and tried to escape across the desert in their lorries, with 11th Hussars in hot pursuit; those who did not surrender were harried until their vehicles were burning wrecks. The survivors, numbering seven officers and 94 men, about one-third of the original force, were herded back to the frontier along with the captured guns and solitary L3, under tow behind a Morris. No British casualties were sustained in the engagement.

Armoured cars continued to play a vital role for both sides throughout the North African campaign, although rarely would their operations be quite as dramatic as those of 11th Hussars during its first weeks. When the front was static most of their time was spent on lonely observation missions far out in the desert, often behind enemy lines, where long periods of boredom and discomfort were punctuated by occasional violent clashes. However, when the situation became fluid, and particularly when an army was in retreat, they were able to exercise their aggressive potential, roaming the enemy's rear areas to destroy transport and snap up prisoners. Between the end of July and the beginning of November 1942, when the contending armies, their flanks covered by the sea in the north and the impassable Qattara Depression in the south, remained entrenched behind thick minebelts, there was little opportunity for armoured car operations. Likewise, in the mountainous areas of Tunisia, where their axes of advance were confined to the valley floors, their activities were again restricted once the front had congealed.

The reconnaissance battalions of the German Army's armoured divisions differed from those of the British both in their composition and the manner in which they were employed. In fact, they incorporated the mechanised counterparts of all those elements which had been present in the cavalry screens of 1914 and consisted of two armoured reconnaissance squadrons with armoured cars; a motor cycle machine-gun squadron; and a heavy squadron containing a troop each of towed 75mm howitzers and anti-tank guns together with an assault pioneer troop with bridging equipment. The armoured reconnaissance squadrons each contained a headquarters troop with one radio command vehicle and four armoured cars fitted with radio; one heavy troop of six SdKfz 231/2 six- or eight-wheeled armoured cars which could be subdivided into three two-car sections, and two light troops each with six SdKfz 222 Series four-wheeled armoured cars. The purpose of the motor cycle machine-gunners and the weapons troops of the heavy squadron was to mount local attacks which would ease the passage of the armoured cars through the lines.

From 1942 onwards the equipment and organisation of the armoured recon-naissance battalions underwent a radical revision, partly because the four-wheeled cars were unable to cope with the Russian mud, and partly because casualties among the motor cycle troops had been heavier than anticipated. Battalions retained six eight-wheeled armoured car troops, progressively re-equipped with versions of the much improved SdKfz 234 Series, which were capable of mounting heavier weapons than the usual 20mm cannon, but the remainder of their subunits were armed with appropriate models of the SdKfz 250 Series of armoured half-tracks as these became available.

Something of the nature of the German armoured cars' work on the Russian Front is described by Colonel Fabian von Bonin von Ostau, who served as a subaltern with the Armoured Reconnaissance Battalion of 1st Panzer Division.

'Having been given a task by division, the commanding officer would despatch several troops along the most important axes and lead them personally. Behind him, the squadron leaders thickened up the screen with further troops. As an officer commanding a section of two eight-wheeled cars, I was given a distant objective, perhaps 20 to 40 kilo-metres into enemy territory, and, without consideration of neighbouring sections, had to reach this using my own initiative. Enemy forces had to be reported and if possible cir-cumvented without detection so that we could penetrate deep into their rear areas. Often we had not reached our objective by nightfall and remained as stationary observers on suit-able features until daybreak. On reaching the objective we were either ordered to return to our unit or were relieved by another section that had followed us up. Occasionally we remained stationary in enemy territory until our own division caught up with us.

'At first one had to overcome and become used to a feeling of loneliness, of being all alone in enemy territory without being able to rely on outside help. With increasing experience, one's self-confidence grew; apart from which, such independent missions were attractive in that one was not pressed into a restrictive framework with one's superiors and neighbours.

'The initial penetration into unknown territory was difficult. For this purpose our own local attacks were taken advantage of before the enemy could recover his bal-ance. When one had achieved some penetration, the advance became easier. A recon-naissance leader must be a good observer and have a nose for knowing where he might run into the enemy. Mostly the cars were well camouflaged and used all available natur-al cover, following each other with the last car covering the rear. On features where a good field of vision was offered, one halted and made a thorough observation. If no enemy were seen then the first car went on to the next observation point, and when it arrived safely the next car was called forward.

'It was important to make a thorough observation of villages as these were nearly always used by the enemy in one form or another. If you see the enemy, then you know. If the enemy is not visible and the civilian population is going about its normal business, then the village is not enemy-occupied. If no people are seen, this is highly suspicious and the village should be bypassed by a wide margin.

'The best patrols I had were those with clean guns. Even worthwhile targets were only reported and not engaged; that is the business of others. A troop leader with a tendency to bang away is useless for reconnaissance purposes since he is soon located by the enemy and chased like a rabbit. A report giving the location of an enemy tank leaguer is of infinitely more value than five shot-up lorries. Reports were made in Morse, as the carrier wave had a greater range than voice transmissions, which were used only between vehicles.'

Armoured reconnaissance in the densely crowded landscape of western Europe presented many more problems than it did in the unconfined space of the North African Desert or the Russian heartland. In September 1944 the Allies launched Operation 'Market Garden' with the intention of ending the war against Germany before the year was out. The operation consisted of two major phases, in the first of which Lieutenant General Sir Frederick Browning's I Airborne Corps was to be dropped at intervals between the existing front line on the Dutch-Belgian frontier and Arnhem on the north bank of the Lower Rhine, the US 101st Airborne Division (Major General Maxwell D. Taylor) north of Eindhoven, the US 82nd Airborne Division (Major General James Gavin) between Grave and Nijmegen, and the British 1st Airborne Division (Major General Roy Urquhart) at Arnhem. The second phase involved Lieutenant General Brian Horrocks' British XXX Corps advancing along the corridor so formed, relieving each of the airborne divisions in turn, and maintaining its progress as far as the Zuider Zee. The effect would be to place the Allies beyond the Rhine and simultaneously outflank both the Siegfried Line and the heavily industrialised Ruhr basin, both of which would be fiercely defended. It would then be possible to drive straight into the heart of northern Germany and secure a victory before the Soviet Army could penetrate Central Europe, saving many thousands of lives in the process.

When the drops took place on 17 September the two American divisions succeeded in capturing their objectives. However, for reasons beyond its control, the British 1st Airborne Division was unable to capture the Arnhem road and rail bridges and was forced back into a perimeter around the suburb of Oosterbeek, where it continued to fight a desperate battle for survival against mounting odds. Meanwhile, XXX Corps, spearheaded by Major General Allan Adair's Guards Armoured Division, reached the 101st Airborne during the afternoon of 18 September and broke through to the 82nd Airborne the following morning. Unfortunately, the advance was halted by tenacious resistance at Nijmegen and XXX Corps began to fall seriously behind schedule. This was not helped by the entire Corps having to advance along a single heavily congested road; nor by the fact that the corridor which had been secured was, in places, little more than a few hundred yards wide, was under constant artillery fire and air attack, and was occasionally cut by the enemy's counter-attacks.

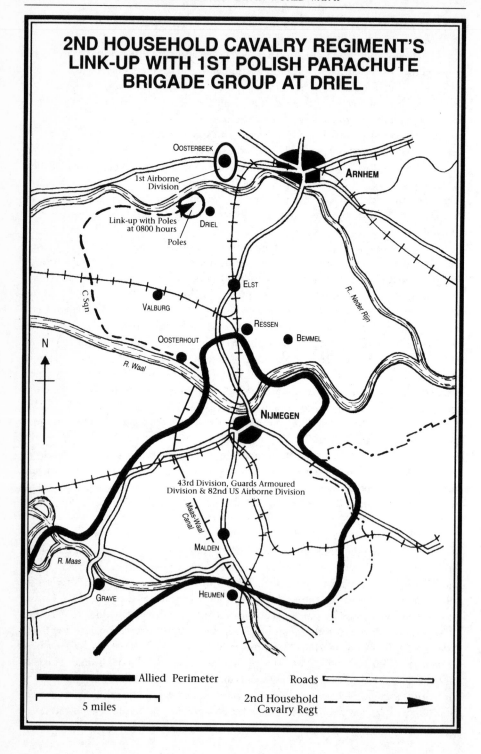

2ND HOUSEHOLD CAVALRY REGIMENT'S LINK-UP WITH 1ST POLISH PARACHUTE BRIGADE GROUP AT DRIEL

OOSTERBEEK

ARNHEM

1st Airborne
Division

Link-up with Poles
at 0800 hours

DRIEL

Poles

C Sqn

ELST

VALBURG

R. Neder Rijn

RESSEN

OOSTERHOUT

BEMMEL

N

R. Waal

NIJMEGEN

43rd Division, Guards Armoured
Division & 82nd US Airborne Division

Maas-Waal
Canal

MALDEN

R. Maas

GRAVE

HEUMEN

Allied Perimeter

Roads

5 miles

2nd Household
Cavalry Regt

Throughout 19 September the Guards Armoured and 82nd Airborne fought their way into Nijmegen, simultaneously fending off counter-attacks from the Reichswald Forest to the east. The battle continued next morning, while an audacious plan was formulated to secure crossings of the Waal and break through to the embattled 1st Airborne at Arnhem. At 1500 Major Julian Cook's US III/504th Parachute Infantry embarked on an apparently suicidal crossing of the river in assault boats. Against all probability, the paratroopers suppressed the opposition on the far bank, captured the Nijmegen rail bridge and were quickly reinforced. At about 1820 a Sherman troop of 2 Grenadier Guards, led by Sergeant Peter Robinson, charged across the Nijmegen road bridge under heavy fire, routed the defenders at the far end at the cost of two of its number, and went on join the Americans; unknown to the Grenadiers, the enemy had attempted to fire the bridge's demolition charges while they were crossing but, for reasons that have never been fully explained, they failed to explode.

On 21 September a squadron of 2 Irish Guards broke out of the Nijmegen bridgehead and pushed forward up the main highway to Arnhem, now only eleven miles distant. Unfortunately, the road ran along the top of a causeway with deep water-filled ditches on either side, so that when the leading Shermans were knocked out by anti-tank guns south of Elst the remainder were unable to deploy; simultaneously, radio communications with the supporting Typhoon ground attack aircraft failed. Behind the burning tanks the rest of the Irish Guards were strung out along the embankment, and behind them were several miles of closely packed vehicles stretched back to the road bridge, presenting an excellent target of which the German artillery took full advantage. It began to look as though the drive to Arnhem had finally stalled until, that afternoon, Major General Stanislas Sosabowski's 1st Polish Parachute Brigade was dropped at Driel, on the south bank of the Lower Rhine opposite Oosterbeek, with the object of opening communications with 1st Airborne Division across the river.

For XXX Corps the priority now became establishing a physical link with the Poles and this became the responsibility of Major General G. I. Thomas's 43rd (Wessex) Division, which had just arrived to take over from the Guards at Nijmegen. For the task itself Adair's armoured car unit, 2nd Household Cavalry Regiment, commanded by Lieutenant Colonel Abel Smith, was transferred to Thomas's command and ordered to provide a squadron which would advance at first light and open a route to Driel.

The regiment's C Squadron was detailed for the task. Major Peter Herbert, the squadron leader, decided to probe the eastern and western shoulders of the bridgehead with two troops, intending to pass the rest of the squadron through any gap that was found. To the east, Lieutenant Corbett's troop soon found itself held up by anti-tank guns, but to the west Captain Richard Wrottesley's troop achieved a clean breakthrough.

'The first mile and a half after crossing the Nijmegen bridge was parallel to the river and on a narrow road banked six feet high on either side. If you want to present yourself as a target there's nothing like having to get nose to tail well silhouetted against the skyline on a road where at no place could you begin to turn round,' wrote one of the squadron's officers who attempted to use the route later in the day. After leaving the embankment, the road passed through the village of Oosterhout, which was known to be in enemy hands, and then the route swung north along minor country roads towards Driel.

141

Fortunately, the same dense morning mist that had inhibited Allied air operations throughout Market Garden now favoured Wrottesley's troop. With visibility down to 50 yards or less, his two Daimler armoured cars and two scout cars moved very quietly along the embankment and through the village, their engines ticking over barely above idling speed. German voices could be heard on all sides through the mist, but, as Wrottesley commented, 'We reckoned that they were as blind as we were, certainly could not man their weapons and, if they heard us, probably thought that we were friendly vehicles – anyhow, the desperate situation of 1st Airborne warranted corresponding risks on our side.'

Map reading presented more difficulties than the enemy, only one of whom was encountered, and he was allowed to fade into the fog since the sound of shooting might have attracted unwelcome attention. By 0800 on 22 September the troop had driven clean through the enemy's lines and was receiving a delighted welcome from the Poles, who were short of anti-tank weapons and under pressure from Elst to the south and Arnhem to the east. Wrottesley immediately informed Herbert of the situation and the latter promptly despatched Lieutenant Arthur Young's troop to join him before the mist could lift. After passing through Oosterhout, Young took a wider loop to the west.

'Things were exceptionally quiet at this time,' he recalled. 'We passed several PzKw IVs which we liked to presume had been knocked out, but these we discovered later from Peter Herbert proved to be very much alive!'

On the way, he picked up several airmen who had been shot down near Arnhem and carried them on to Driel. Shortly after, Herbert tried to get through with the rest of the squadron but by then the mist had lifted and the advance was halted by several tanks firing straight down the embankment from Oosterhout. After the leading scout car was knocked out, Corporal of Horse Brown worked his way round it and punched an armour piercing round through an outhouse behind which a PzKw III was firing from a partially concealed position; later in the day, the tank was found to have been penetrated and abandoned. Brown then put down smoke, blinding the enemy gunners and enabling the squadron to pull back from its exposed position while an infantry attack was mounted.

At Driel the two armoured car troops were kept fully occupied, patrolling the surrounding area and helping to beat off the enemy's counter-attacks. At one stage there was a clash with an enemy patrol resulting in the capture of a large scale map belonging to the river engineer at Driel; this provided details of currents and bank surfaces and proved to be very useful in deciding how contact was to be made with 1st Airborne on the far shore. The Poles were also grateful for the cars' additional firepower, as their own ammunition supply had become dangerously depleted by the end of the day's hard fighting.

In addition, when he was not actually fighting, Wrottesley was providing the only radio link between Driel and XXX Corps, via Herbert. 'God knows how big were the Sunrays (formation commanders) listening in, and chipping in, on the rear link set. For every one message sent back by the cars to Regiment, Corps sent forward two further questions to be answered.' Each of these Wrottesley dealt with calmly, sometimes with a wry touch of humour.

'Has the Arnhem railway bridge been blown?'

'Yes, it has.'

'Who is holding the northern end?'

'By the shots that are being directed at us from that area, I presume it must be the Germans.'

'What about the Heveadorp Ferry?'

'The ferry has been destroyed.'

'Can you see if there are any other ferry boats lying about? What is the state of the tracks leading to the river's edge and will it carry the DUKWs (amphibious lorries) without bogging? The CRE (Commander Royal Engineers) 43 Division wishes to know the width of the gap in the blown railway bridge – can you send back a bridging report?'

And so on.

Simultaneously, when it too was not fighting, Young's troop was observing for the guns of 64 Medium Regiment, Royal Artillery, which provided the hard pressed 1st Airborne with the first tangible support it had received from XXX Corps. Two of Urquhart's staff officers who had crossed the river in a rubber boat also used the Household Cavalry net to report in chilling terms the desperate situation in Oosterbeek: 'We are short of food, ammunition and medical supplies; we cannot hold out for more than twenty-four hours; all we can do is hope and pray.'

In the meantime, Thomas's division had attacked on a two-brigade frontage out of the Nijmegen bridgehead. By evening, 129th Brigade was still making slow progress towards Elst, but 214th Brigade had taken Valburg and Oosterhout. A relief column, consisting of 5th Duke of Cornwall's Light Infantry, a Sherman squadron belonging to 4th/7th Dragoon Guards and two DUKWs laden with supplies, was pushed through the gap and headed for Driel. On the way, a sergeant major at the rear of the column noticed that it was being followed by several Tiger tanks that were maintaining their correct convoy distance. Without giving any hint of alarm, he sent a motor cycle despatch rider to warn the column commander who dropped off several PIAT teams to set up an ambush. When the leading Tiger rounded a bend it was knocked out by a shower of hollow-charge bombs. The second Tiger cannoned into it and the third, taking avoiding action in the half light, careered off the road into a drainage ditch, to be followed in succession by four Panthers. Altogether, the DCLI destroyed five enemy tanks in this ambush.

The location of the Polish perimeter had been transmitted to HQ 43rd Division but, for some reason neither this information, nor the fact that two Household Cavalry troops were present with the Poles, was relayed to the relief column, with tragic consequences. Young was aware that the column was on its way, but could not assume that the approaching sound of tracks was made by friendly vehicles and moved his troop forward, displaying their yellow recognition strips as a precaution. In the gloom these were either unseen or misinterpreted by the leading Sherman, which opened fire, damaging Young's car and killing a scout car driver. It took all Young's powers of persuasion to stop the enraged Poles lynching the mortified tank crew when they arrived.

Early the following morning Wrottesley's troop, escorting one of Urquhart's staff officers back to Nijmegen, became involved in a brief duel with a Panther which had evidently survived the previous evening's ambush and been recovered from the ditch, since it was described as having a 'dirty green nose.' Although one car was lost there were no casualties and the troop delivered its charge safely.

This ended the two troops' critical part in establishing contact with 1st Airborne Division, the remnants of which, just 2163 men of the original 10,000, were evacuated in assault boats during the night of 25/26 September. The 2nd Household Cavalry already had much to its credit, for in Normandy the regiment's discovery of an unguarded bridge lying on the boundary between two German armies had influenced the subsequent course of the battle, and during Market Garden it had been the first unit of XXX Corps to establish contact with the two American airborne divisions. Yet, the breakthrough of Wrottesley's and Young's troops to the Poles, and their activities with them, were to be described by some as the Regiment's 'most brilliant action.' It would indeed be difficult to find another example in which so many of the armoured car's roles had been employed within a comparatively short space of time to produce such important results.

Honour in Adversity
Holding Actions on the Road to Bastogne

n 1939 the Regular Army of the United States, including the Army Air Corps, numbered just 188,565 men. Its tank strength amounted to less than 400 vehicles, many of which were obsolete designs, divided between the Infantry and the Cavalry, although for legal reasons the latter, technically denied tanks, referred to theirs as combat cars. It was not until 10 July 1940, following the German victories in Poland and France, that the establishment of a joint Armored Force put an end to sectional interests.

It was clear that, sooner or later, the United States would become involved in the global conflict. Very quickly the American genius for organisation was harnessed to the country's enormous industrial potential with the result that tanks soon began rolling off the production lines in very large numbers. These included the recently standardised M3 Light Tank (Stuart), which was a good design for its weight, the 75mm-armed M3 Medium (Lee), which had been developed at very short notice from the existing M2 Infantry Tank, and the M4 Medium (Sherman), of which no less than 49,234 were built by June 1945. In response to the demands of the gun-armour spiral, later Sherman models were completed with 76mm guns and the 90mm-armed M26 Pershing, designed as an answer to the German Tiger, entered service in small numbers during the closing months of the war. Simultaneously, huge numbers of self-propelled artillery mountings and half-tracked personnel carriers for armoured infantry units also began leaving the factories.

In total, the US Army fielded a total of sixteen armoured divisions, of which all but one (the 1st, which served in North Africa and then Italy) were committed to the campaign in North West and Central Europe. By 1944 the American armoured division had become a balanced force with what was arguably the most flexible command system in the world. This consisted of three Combat Commands, A, B and Reserve or R (CCA, B and R), to which divisional resources were allocated as the situation demanded. These resources consisted of three tank battalions (six in the case of 2nd and 3rd Armored Divisions), three armoured infantry battalions, three self-propelled artillery battalions, a tank destroyer battalion, an armoured cavalry reconnaissance squadron, an engineer battalion and divisional services. In addition, the Army formed 65 non-divisional tank battalions, of which more than half served in Europe and the rest in the Pacific. These were generally allocated to provide close support for infantry divisions, but sometimes operated as semi-independent armoured groups.

The dramatic victories won by the German armoured corps during the early years of the war were studied at the highest levels, and the conclusion was reached that the problems caused by a mass breakthrough of enemy armour required a radical solution. The towed anti-tank gun was clearly not the answer; on the other hand, the anti-tank gun was potentially a more powerful weapon than that carried by most contemporary tanks and, given self-propelled mobility, it could be deployed rapidly in large numbers in the path and on the flanks of an enemy mass tank attack, where its opponents would be destroyed by direct gunfire. On 14 May 1941 General George C. Marshall, the Army's Chief of Staff, issued a directive establishing what amounted to a completely new branch of service, the Tank Destroyer Force (TDF). The basic unit of this was the battalion although, as the underlying intention was employment en masse, three battalions would constitute a Tank Destroyer Group and two such groups would form a Tank Destroyer Brigade.

The first equipment issued to the TDF consisted of the legendary if now somewhat elderly French Seventy-Five, several hundred of which were held in store, mounted on M3 half-tracks. This was followed in 1942 by the M10 tank destroyer, based on the M4A2 medium tank chassis, fitted with an open-topped turret housing a three-inch high velocity gun. However, the appearance of larger, better armed German tanks early in 1943 meant that the TDF required an even more powerful gun to achieve penetration and a new turret housing a 90mm gun was fitted to the M10 chassis, the new vehicle being standardised as the M36 in June 1944. Simultaneously, the TDF was designing its own fast tank destroyer, the M18, armed with a 76mm gun, and this entered service in the autumn of 1943.

It was ironic that by the time American troops took the field against the German Army, not only were the days of the latter's armour-led breakthroughs all but over, but also that, because of the influences of terrain and the development of new weapons such as the bazooka, the tactical picture on the battlefield had itself undergone a radical change. Nevertheless, there were occasions during the fighting in Tunisia when tank destroyer battalions were employed in accordance with their training and scored heavily. Subsequently, they found themselves employed in a variety of roles, including anti-tank defence, supplementary artillery and close gunfire support. Each armoured division possessed its attached tank destroyer battalion and some infantry divisions had two.

The Armored Force did not fight its first major action until February 1943, when the inexperienced 1st Armored Division was worsted at Kasserine Pass in Tunisia, but managed to roll with the punch and come back fighting. Thereafter, with the exception of the short campaign in Sicily, the Force was unable to demonstrate its real potential for fully eighteen months because of the conditions under which the Italian and Normandy campaigns were fought.

Nevertheless, it is certainly worth mentioning in passing that the latter gave rise to one of the most ingenious inventions of the war. The strategy of the campaign, it will be recalled, involved an almost continuous series of offensives by the British Second Army, intended to pin down the German armour while the American First Army prepared to break out of the beachhead to the south. However, this process also involved the Americans expanding their own perimeter in order to accommodate the steady flow of troops arriving in Normandy. The bocage country, consisting of small fields surrounded by steep earth banks topped with hedges, was as heartily disliked by American tank crews as it was by the British, since it not only provided the enemy with dense cover but was also an obstacle. Inevitably, whenever a bank was climbed the tank's nose would rise, exposing the vulnerable belly plates to an anti-tank round or a panzerfaust, with fatal results. The complete answer was provided by Sergeant Curtis C. Culin, Jr, of the 102nd Cavalry Reconnaissance Squadron, who welded several iron spikes to the nose of a Stuart. These dug into the bank, so preventing the nose rising, and the tank simply pushed its way through. The effect was to reduce American casualties and cause some consternation among the enemy, who lost much of the advantage of the favourable defensive terrain. Known as the Culin Prong or Rhino, the device was welcomed with such enthusiasm that when the American breakout, codenamed 'Cobra', commenced on 25 July 1944, three out of every five tanks involved were fitted with it, as were a large number of tanks on the British sector, using scrap iron salvaged from former German beach defences. It is pleasing to note that Sergeant Culin was rewarded with membership of the American Legion of Merit.

In the immediate aftermath of Cobra, the US Third Army was activated on 1 August under the command of Lieutenant General George S. Patton. After clearing the Brittany peninsula, this swung east, trapping the German armies in Normandy within a pocket from which few escaped. Throughout the remainder of August and on into September Patton drove his army across northern France until, as he approached the German frontier, the strain on his logistic resources became too great and, lacking the fuel to advance further, he was forced to halt. By the time the situation was restored, poor weather had begun to restrict the amount of air support he could receive and the Germans, having rallied after their defeat in France, were fighting tenaciously in defence of their homeland.

Patton's dashing style of warfare was almost certainly in the mind of General Dwight D. Eisenhower, the Allied Supreme Commander, when he was asked by journalists to comment on the relative characteristics of the British and American troops under his command. He answered that for prolonged operations involving continuous fighting he would prefer to employ British troops, but for fast moving operations his choice would rest with Americans. The British, of course, had already provided ample evidence that they were capable of conducting mobile warfare, just as the Americans had proved they were capable of dogged fighting, but Eisenhower's remark did contain a basic core of truth. In fact, the real worth of any army is only revealed in adversity, and during the so-called Battle of the Bulge that of the US Army became apparent when it was simultaneously required to conduct a stubborn defence at several points while reacting at speed to eliminate a potentially dangerous strategic threat.

Adolf Hitler, sick in mind and body, his nerves in tatters after the failed July Bomb Plot, was more than ever a prey to grand delusions. Perceiving some imagined link between himself and Frederick the Great, he informed his generals that just as the latter had taken the offensive and won remarkable victories when his fortunes were at their lowest ebb, so too would he. The Ardennes had served him well in 1940, and he intended that they should do so again in 1944, providing a gateway for a great offensive drive directed across the Meuse and on to Antwerp, the capture of which would isolate the Canadian First, British Second and US First and Ninth Armies in a pocket to the north. Enclosed in a ring of steel, he reasoned, they would be faced with the alternatives of destruction or a second Dunkirk, and in such circumstances the Western Allies would quickly recognise that victory was impossible and sue for a negotiated peace.

Most German commanders had long since recognised the dangerous futility of arguing with the Führer. It was pointless to stress that the conditions prevailing in 1940 did not exist in 1944; or that whereas the scanty Ardennes road system had favoured the south-westerly axis of advance taken in 1940, the projected offensive had a north-westerly axis with even fewer roads at its disposal and ran across the grain of the country; or

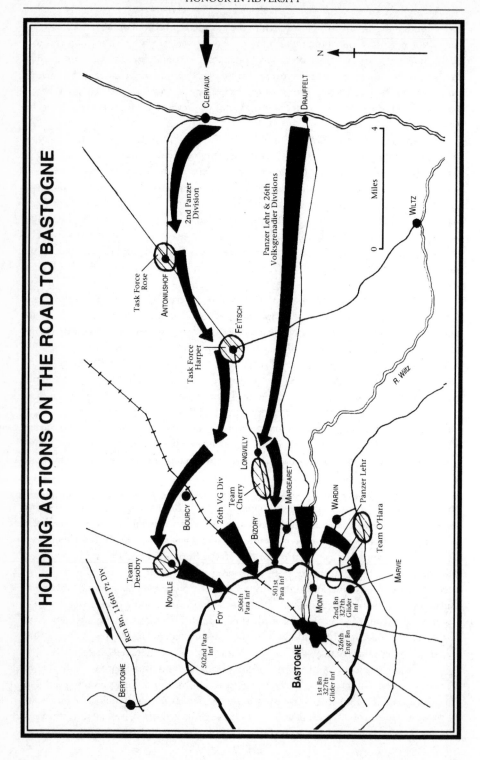

HOLDING ACTIONS ON THE ROAD TO BASTOGNE

2nd Panzer Division

Panzer Lehr & 26th Volksgrenadier Divisions

N

Miles

4

0

CLERVAUX

DRAUFFELT

WILTZ

Task Force Rose

ANTONIUSHOF

Task Force Harper

FEITSCH

R. Wiltz

LONGVILLY

Team Cherry

26th VG Div

BOURCY

MARGEARET

WARDIN

Panzer Lehr

BIZORY

Team O'Hara

MARVIE

Team Desobry

Rcn Bn, 116th Pz Div

NOVILLE

501st Para Inf

MONT

2nd Bn 327th Glider Inf

BERTOGNE

FOY

506th Para Inf

502nd Para Inf

BASTOGNE

326th Engr Bn

1st Bn 327th Glider Inf

that fuel stocks were 75 per cent below requirements for the operation, an argument countered with the optimistic prediction that captured American stocks would prove to be adequate. The private reactions of Field Marshal Gerd von Rundstedt, the recently reinstated Commander in Chief West, and Field Marshal Walter Model, whose Army Group B would carry out the operation, were contemptuous, but further discussion was not permitted.

Germany's entire strategic reserve was therefore concentrated in great secrecy to the area of the sector chosen for the attack, lying between Monschau in the north and Echternach in the south, a distance of 85 miles. On the right was SS General Sepp Dietrich's Sixth SS Panzer Army; in the centre was General Hasso von Manteuffel's Fifth Panzer Army; and on the left was General Erich Brandenberger's Seventh Army, which had the responsibility of protecting the southern shoulder of the penetration.

By coincidence, the Ardennes was the weakest sector of Allied line, just as it had been in 1940. To the men of Lieutenant General Courtney Hodges' US First Army it was known as the Ghost Front, for the simple reason that nothing ever seemed to happen there – it was, in fact, just the right sort of area in which tired divisions could be rested and recently arrived divisions given their first, small-scale introductions to battle. From Monschau to the Losheim Gap the front was the responsibility of Major General Leonard Gerow's V Corps, two of whose divisions, the 2nd and 99th, together with CCB 9th Armored Division, lay directly in the path of the offensive, the rest of the corps being still engaged in offensive operations to the north. From the Losheim Gap south to Echternach, a distance of over 60 miles, the front was held by the 28th and 106th Infantry Divisions of Major General Troy Middleton's VIII Corps with CCA 9th Armored in support. Covering the boundary between the two corps, and indeed the Losheim Gap itself, was a single armoured car squadron of the 14th Cavalry Group.

There had, of course, been indications that the enemy was up to something, but these were misinterpreted by higher authority. Therefore, when 2000 guns opened fire along the Ghost Front at 0530 on 16 December, the Germans achieved total strategic and tactical surprise. In the north their infantry attack was stalled by the stubborn defence of the US 2nd and 99th Divisions, with the result that the 12th SS Panzer Division Hitlerjugend was also unable to advance. In the Losheim Gap the armoured cars of the 14th Cavalry Group were too few in number and too lightly equipped to impose more than local checks and report the enemy's progress, and here a breakthrough was achieved by Battlegroup Peiper, the spearhead of 1st SS Panzer Division Leibstandarte Adolf Hitler. Nevertheless, by the end of the day the Sixth SS Panzer Army was several hours behind its schedule.

On the sector of Manteuffel's Fifth Panzer Army the Germans were doing rather better. Two regiments of the completely inexperienced US 106th Division were isolated on the Schnee Eifel and eventually forced to surrender. At 1600 the German armour was called forward and began to advance into what had been the American rear areas by the light of searchlights reflected from the clouds.

At this stage senior American commanders were clearly unable to determine the real object of the German offensive. It was, however, apparent that it was a major strategic undertaking. The general belief was that Rundstedt, a skilled and respected professional, was directing events, and for this reason alone it had to be taken very seriously;

few could have guessed that Rundstedt was taking as little interest as possible in the exe-cution of Hitler's plan, or that the controlling mind was that of Hitler himself. At the lower command levels American formation and unit commanders struggled to cope with local situations that were deteriorating by the hour. Some men proved unequal to the task, among them the commanding officer of the 14th Cavalry Group, who had to be sent to the rear in disgrace. Others, shocked by the weight of the German attack, began streaming back from the front when the command links were broken, and in so doing blocked the roads along which reinforcements were trying to reach it. In places the con-fusion quickly became chaos when English-speaking Germans in American uniforms, commanded by SS Colonel Otto 'Scarface' Skorzeny, began misdirecting traffic or spreading panic.

And yet, amid the snow-covered, gently rolling, wooded hills of the eastern Ardennes, there were many more who stood their ground, forming impromptu battle-groups around a few Shermans, tank destroyers and anti-tank guns. Knowing nothing of the purpose of the German offensive, they understood instinctively that any delay they could impose, whether it was a road blocked by burning enemy tanks or a bridge blown at the last minute, would not only cost their opponents priceless time while they sought alternative routes, but also buy time in which their own commanders could coordinate a response. That night Eisenhower ordered two armoured divisions, the 10th from Patton's Third Army to the south and the 7th from Lieutenant General William H. Simpson's Ninth Army to the north, to move into the Ardennes, followed by his strategic reserve, Major General Matthew B. Ridgeway's XVIII Airborne Corps.

The Germans received their first serious check at the ancient town of Clervaux, which contained the headquarters of the 110th Infantry, part of Major General Norman D. Cota's 28th Division. Colonel Hurley Fuller, commanding the 110th, also had two companies of the division's 707th Tank Battalion at his disposal and these he was forced to use piecemeal to shore up his line. One Sherman platoon, for example, reached the nearby village of Bockholz just in time to prevent an artillery battery from being overrun, recapturing a gun that had been temporarily lost.

The following morning Fuller was informed that a force of German tanks and panzergrenadiers were approaching Clervaux from the direction of Marnach, a village to the east where one of the 110th's companies had already been overrun. He sent his five-strong reserve tank platoon up onto the high ground surrounding the town, where it destroyed four PzKw IVs at the cost of three Shermans and temporarily halted the Ger-man advance. Unfortunately, while returning to Clervaux the two surviving Shermans took a wrong turning and were not seen again. Fuller therefore called in another Sher-man platoon from one of his outlying companies and as this approached the outskirts of the town it destroyed a further PzKw IV, the wreckage of which effectively blocked the route from Marnach. After this, the German armour desisted in its attempts to force its way into Clervaux and simply fired into the town from the surrounding heights. Cota sent up a platoon of tank destroyers from CCR 9th Armored, hoping that these would enable Fuller to deal with the threat, but after exchanging a few shots with the enemy they retired in such haste that one overturned on a hairpin bend. Panzergrenadiers now began to infiltrate the town from several directions. At length, faced with overwhelming numbers, Fuller and his staff were compelled to escape through the rear of their command post in the Claravallis Hotel, only to be captured as they sought to regain their own lines. The 110th's Headquarters Company continued to resist in the medieval castle, the stout walls of which became pitted with tank and artillery shells, denying the route through Clervaux to all but armoured vehicles; they did not surrender until the following morning, by which time parts of the building were in flames and a German tank had rammed its way through the main gate. Elsewhere, in the surrounding villages, companies of the 110th Infantry and 707th Tank Battalion continued to fight on until by 20 December their respective losses amounted to 2700 men and some 60 tanks.

Major General Troy Middleton, commanding VIII Corps, was acutely conscious that the fall of Clervaux had opened the road to Bastogne, the hub of a road network that the enemy would need to capture in order to maintain the westerly direction of his offensive, and he immediately ordered CCR 9th Armored Division to establish two roadblocks along the route. The first of these was situated to cover a road junction at Antoniushof and the second covered another road junction at Fe'itsch, some four miles to the rear and only seven miles from Bastogne itself.

The Antoniushof block was manned by Task Force Rose, consisting of a Sherman company, a company of armoured infantry, a tank destroyer platoon and an engineer platoon with an artillery battery in support, under the command of Captain Lawrence K. Rose; nearby, although not under command, was a company of 110th Infantry. During the afternoon of 18 December the position was attacked by Panthers and PzKw IVs of the 2nd Panzer Division, supported by panzergrenadiers. Outgunned, seven of the Shermans were knocked out and, seeing this, the armoured infantry decamped, followed by the artillery, whose positions were coming under small arms fire. When Middleton refused to grant Rose's request to fall back on Fe'itsch, the latter held on until dusk then broke out to the north-west with his remaining tanks and tank destroyers, making for Houffalize. Unfortunately, the survivors ran into the 116th Panzer Division moving westwards and few of them escaped. Through the gathering darkness, 2nd Panzer rolled on towards Fe'itsch and Bastogne.

The Fe'itsch block was held by Task Force Harper, consisting of a Sherman company and an armoured infantry company with two artillery batteries in support, under the command of Lieutenant Colonel Ralph S. Harper. Like the rest of 9th Armored Division, the task force was inexperienced and became unsettled when, following the outbreak of violent firing in the direction of Antoniushof, the demoralised infantry and artillerymen of Task Force Rose began streaming in with tales of the enemy's superior firepower. Nevertheless, Harper's troops settled into their positions, confident in the accepted contemporary wisdom that a major tank attack was unlikely during the hours of darkness. That, however, was not the case. A number of 2nd Panzer Division's Panthers had been fitted with infra-red light projectors and now, for the first time, these devices were employed on the battlefield. The Americans suddenly found themselves illuminated without being able to locate the source of the light. Very quickly, ten Shermans were set ablaze, Harper was killed, and the panzergrenadiers closed in to rake the whole area with fire. According to German sources, only three tanks offered serious opposition. The rest of the task force broke and streamed back down the road to the next village, Longvilly, where they were rallied by the headquarters personnel of CCR 9th Armored.

During the evening the first of the reinforcements promised by Eisenhower, CCB 10th Armored Division, had driven into Bastogne under the command of Colonel William Roberts. At Middleton's insistence Roberts divided CCB into three teams, one of which, designated Team Cherry after its commander, Lieutenant Colonel Henry T. Cherry, was promptly despatched to Longvilly. Cherry, who had at his disposal a Sherman company, two Stuart platoons, a company of armoured infantry and a few armoured cars, reached Longvilly at about 1900. At that time Task Force Harper was still in position, but the village was crowded with vehicles belonging to CCR 9th Armored and he decided to halt his column to the west. After directing his advance guard, composed of Stuarts and armoured cars under Lieutenant Edward Hyduke, to occupy a position on high ground beside the highway, he went into the village to discuss the situation with Colonel Joseph Gilbreth, the commander of CCR. Neither officer saw any cause for immediate concern and Cherry returned to Bastogne to report to Roberts. On the way, while passing through the village of Magaret, shots cracked past his jeep but, believing that they had been fired by trigger-happy stragglers, he carried on.

Unknown to Cherry, or to anyone else for that matter, 2nd Panzer's running mate, the Panzer Lehr Division, had been advancing along a parallel axis to the south and it had just cut the Bastogne road at Magaret. On his return journey, therefore, Cherry found German infantry and armour in possession of the village. By radio, he ordered Captain William F. Ryerson, commanding the team's armoured infantry company, to clear a way through, leaving Hyduke in position to act as rear guard. Because of the congestion this was far easier said than done, but at length Ryerson got his Shermans and half-tracks turned round and on the move. However, as it emerged from a cutting near the village, the leading tank was knocked out and set ablaze. As the infantry dismounted to deploy for attack two anti-aircraft half-tracks, hell-bent on reaching safety, careered through the cutting and crashed at speed into the wreck, creating a tangle that completely blocked the road.

In the meantime, Task Force Harper had given way and its shaken remnants had reached Longvilly. Shortly after midnight, Gilbreth decided to send his headquarters troops back to Bastogne, unaware of the ugly situation that had developed in his rear. Seeing the withdrawal commence, the demoralised survivors of the road blocks believed that the order applied to them as well. A scramble to get out developed into a panic that was not brought under control until CCR's vehicles had become jammed solidly with those of Team Cherry between Longvilly and Magaret.

It took all the next day, 19 December, for Team Cherry to fight its way through Magaret. During that time the trapped column, unable to leave the road because of soft going on one side and a steep slope on the other, came under fire from artillery and tanks and vehicles began to burn along its length. Particularly troublesome was the long-range shooting of Panzer Lehr's tank destroyer battalion, positioned on a ridge south-west of the highway. At length some vehicles managed to escape along a side road leading north while others took a dirt track leading west to Bastogne from the outskirts of Magaret. Back in Longvilly, Hyduke, who was to be killed in Bastogne, had the fight of his life, his small rearguard being simultaneously engaged by elements of 2nd Panzer, Panzer Lehr and the approaching 26th Volksgrenadier Division, one the best in the German service. He supplemented the limited firepower at his disposal by persuading the last surviving platoon of CCR's 811th Tank Destroyer Battalion to join him and knocked out eight of his opponents' tanks. In the process, he inevitably sustained losses of his own and after an hour's intense fighting Cherry ordered him to disengage. Since it was clearly impossible to use the road, he was forced to destroy his remaining vehicles and lead his men out on foot. Team Cherry's losses during the operation amounted to 175 men, ten medium and seven light tanks, and seventeen half-tracks.

In the meantime, Roberts had despatched his second team, consisting of an armoured infantry battalion and a Sherman company under the command of Lieutenant Colonel James O'Hara, to Wardin, where it covered the main approach road to Bastogne from the south-east. In contrast to Cherry's men, Team O'Hara spent a quiet night, but from midday onwards on 19 December it began trading losses with Panzer Lehr and after dusk was pulled back to the high ground overlooking Marvie, where it constructed a roadblock with logs.

Roberts' third team, made up of fifteen Shermans, an armoured infantry company and some armoured cars under the command of Major William Desobry, was sent north from Bastogne to Noville, a village where three roads met. Arriving in position at 0230 on 19 December, Desobry set outposts and ninety minutes later one of these ambushed a group of enemy reconnaissance half-tracks, approaching from the east. After an exchange of grenades, both sides withdrew.

The night was misty and shortly after this incident the dull roar of many engines penetrated the darkness, gaining in strength and apparently bypassing the village to the north. At about 0600 the indistinct shapes of two German tanks emerged through the fog from this direction, moving south along the road from Houffalize to Bastogne. As stragglers had been passing through their position for some time, the Americans held their fire until it was too late, with the result that two of the four Shermans present were set ablaze. Desobry withdrew the outpost into his perimeter. With the coming of partial light, the two German tanks renewed their advance, this time supported by panzergrenadiers. They ran into a storm of fire and were both knocked out, partially blocking the highway, and their infantry scattered into the fog.

There were no further developments until approximately 1030 when the mist suddenly lifted, enabling Desobry to take stock of his position. It was hardly enviable, for Noville was dominated by ridges to the north and north-east. Fourteen German tanks could be seen lined up along the northern ridge, while to the north-east and elsewhere another fifty or sixty were visible. A fire fight started at once, and in this the Americans, now reinforced by a platoon of the 609th Tank Destroyer Battalion, had a decided advantage in that they were partially concealed and protected by the buildings in the village, whereas their opponents were fighting in the open. By the time it was over, the 2nd Panzer Division had lost a further seventeen tanks; Desobry's loss amounted to one tank destroyer and several smaller vehicles destroyed, plus seventeen men wounded. Incredibly, one Panther fell victim to the 37mm gun of an M8 armoured car. In theory, this could be achieved either by deflecting a round from the underside of the curved mantlet through the thinner roof armour of the driving compartment, or by striking the ground immediately ahead of the vehicle so that it ricocheted upwards through the belly plates, although the chances of achieving such a strike were remote; whatever the point of aim, the kill astonished the 37mm gunner as much as it did the Panther crew.

By now Desobry was beginning to feel that his team was far out on a limb, but he was informed that reinforcements were on the way. The leading units of Brigadier General Anthony McAuliffe's 101st Airborne Division had reached Bastogne and during the afternoon the I/506th Parachute Infantry, commanded by Lieutenant Colonel James LaPrade, marched into Noville. Together, Team Desobry and the paratroopers counter-attacked in an attempt to capture the high ground but were beaten back. However, during the evening a platoon of five M18 tank destroyers belonging to the 705th Tank Destroyer Battalion was sent forward to join them. Throughout the night the defences were probed in twos and threes by the German armour and both sides sustained casualties in close quarter fighting. Among them were LaPrade, killed, and Desobry, seriously wounded and captured before he could be evacuated; command of Team Desobry and the I/506th was assumed, respectively, by Major Charles Hustead and Major Robert Harwick.

On 20 December the fog suddenly lifted in mid-morning, just as it had done the previous day. The 705th's M18s, armed with 76mm guns, went into action at once against fifteen German tanks on high ground to the east. Four of these were knocked out with the first shots and the rest retired behind the crest. Against this success, the clearer visibility also revealed that the Germans had cut the road between Noville and Bastogne. At 1300 McAuliffe, to whom CCB had now been formally attached, ordered the two majors to fight their way out, assisted by a paratroop battalion that was to attack the intervening village of Foy. By the time the necessary preparations had been made the fog had fortunately descended again, enabling most of the two units to get through to Bastogne at dusk; three German tanks attempted to interfere with the withdrawal and destroyed four Shermans, but were driven off after one of their number was knocked out by a tank destroyer.

In total, Team Desobry's losses amounted to eleven tanks, five tank destroyers and approximately 200 men; the I/506th sustained the loss of 212 men, killed, wounded and missing. The losses imposed on 2nd Panzer Division included thirty tanks and an estimated 600 personnel casualties, plus a delay of two days that could never be made good.

Team O'Hara was also in action again on 20 December. After his reconnaissance unit had failed to penetrate its roadblock, Lieutenant General Fritz Bayerlein, commanding the Panzer Lehr Division, decided that by striking through Marvie, three-quarters of a mile to the west, he could regain the main Bastogne road some distance to the north of the village. He was unaware that Marvie now lay within the perimeter that McAuliffe was forming around the town with 101st Airborne Division, or that the village's scratch garrison of weary engineers had been replaced by II/327th Glider Infantry, supported by Team O'Hara's Stuarts; and he certainly did not suspect that, situated on high ground to the north-east and on the flank of his projected attack were two of the team's Shermans.

Following a brief preparatory bombardment of the village, the Panzer Lehr battlegroup, consisting of four PzKw IVs, an assault gun and six panzer grenadier half-tracks, broke cover at 1125 and advanced across open country. In an exchange of shots two of the Stuarts in Marvie were destroyed and, with the approval of the airborne commander, the remainder withdrew. Thus encouraged, the battlegroup began closing in rapidly. At a range of only 700 yards it presented the two Shermans with broadside targets; in quick succession two PzKw IVs and a half-track were knocked out. A second half-track bogged down and was hastily abandoned before it too was smashed by an armour piercing round, and the third PzKw IV retired into cover. As the rest of the battlegroup broke into Marvie, the assault gun shot up several parked vehicles but was then hit by both Shermans and burst into flames, and the last PzKw IV fell victim to an infantry bazooka. A sharp battle between the panzergrenadiers and the glider infantrymen left 30 of the former dead and a similar number surrendered. In addition to the two light tanks lost, American casualties amounted to five killed and fifteen wounded.

These continual checks had a profound effect on the German conduct of operations. Manteuffel's Fifth Panzer Army, the spearheads of which should have been approaching the Meuse on the second evening of the offensive, was now days behind schedule. General Heinrich Freiherr von Luttwitz, commanding XLVII Panzer Corps, felt reasonably certain that, using 2nd Panzer, Panzer Lehr and 26th Volksgrenadier Divisions, he could capture Bastogne quickly. Manteuffel, however, mindful of the overall objective, ordered him to continue his drive towards the Meuse. 2nd Panzer and Panzer Lehr therefore bypassed the town respectively to the north and south, leaving it to be attacked by the 26th Volksgrenadiers and other infantry formations. By now, however, it must have been apparent even to the most optimistic of the German commanders that the Americans had recovered from their initial shock and that the critical opening phase of the offensive had failed to produce the results Hitler had predicted.

The Americans themselves were well aware that in committing their armour in small groups during this phase they were acting contrary to established principles and would incur heavy losses because of it. Yet, no other alternative existed and the cumulative effects of each comparatively minor action produced the desired result. The stand of 110th Infantry and 707th Tank Battalion in the area of Clervaux bought sufficient time for Task Forces Rose and Harper to move into position; in turn, these bought just enough time for Teams Cherry, O'Hara and Desobry to be deployed on the principal routes into Bastogne; and while these last were fighting their battles, the 101st Airborne Division was consolidating its hold on the town itself. It had been a very close-run thing, with a wafer thin margin between success and failure.

Sadly, considerations of space must leave untold the stories of 101st Airborne Division's epic defence of Bastogne and of its relief by 4th Armored Division on 26 December; of the equally stubborn defence of St Vith by Brigadier General Robert Hasbrouck's 7th Armored Division; of Battlegroup Peiper's massacre-strewn journey along the Amblève valley in an abortive attempt to capture the fuel dumps which alone would permit the offensive to continue; or of the defeat of 2nd Panzer and Panzer Lehr, their fuel tanks almost run dry, by the US 2nd Armored Division and the British 29th Armoured Brigade at Celles near the Meuse, after which Manteuffel ordered the survivors to withdraw.

The last traces of the 'Bulge' created by the German offensive were finally eliminated in January. The month-long campaign in the Ardennes cost each side about 800 tanks; the Germans sustained 100,000 casualties, the Americans 81,000 and the British 1400. The Western Allies' plans had been set back by approximately six weeks, during which the Soviet Army made enormous territorial gains in the east.

The Japanese Dimension

ven today, many people are unaware of the full extent to which tanks played a critical role in the Far Eastern campaigns of World War II. With some justice, these are regarded as being primarily an infantry war, fought in the jungles of Burma, the Solomon Islands and Papua/New Guinea and among the islands of the Central and South Pacific where, superficially at least, tanks had little or no place. Yet, no fewer than twelve British and Indian armoured regiments were employed in Burma; the US Marine Corps' divisions, which took part in the majority of amphibious operations, contained their own organic tank battalions, while the US Army committed one third of its independent tank battalions to the Pacific Theatre, plus tank destroyer and other armoured units; and Australian armoured regiments fought in Papua/New Guinea, Borneo and elsewhere. It is true that in the majority of cases tanks fought in support of infantry to secure limited objectives and in these circumstances it was inevitable that, to some extent, their actions should be obscured by those of the higher formation to which they were attached. Nevertheless, during the 1945 fighting in Central Burma, which possessed more open going, British and Indian armour was handled very much according to the principles of Fuller's Plan 1919; the enemy's vital communications centre of Meiktila was captured by coup de main, his rear areas were ravaged by all-arms columns, his main front along the Irrawaddy was broken by sustained pressure and, finally, he was harried without mercy all the way to Rangoon.

Fighting the Japanese in Burma or the Pacific presented Allied tank crews with problems that were not encountered elsewhere. This was not because Japanese tanks were in any way formidable opponents, for those most frequently encountered, the Type 95 Light and the Type 97 Medium, were undergunned, thin skinned and of obsolete design; nor that Japanese armoured corps, tied by tradition to the role of infantry support, possessed a highly developed instinct for the tank battle, which it sensibly avoided whenever possible; rather, it was the nature of the Japanese Army itself, whose members possessed an unshakeable belief that their lives were already forfeit to the Emperor and that the ultimate sacrifice was simply the final act in a lifelong process of religious and philosophical dedication.

Of course, Japanese tanks did have their successes, notably against the tankless British army in Malaya, where they were used to fillet their way through the road-bound defences while their infantry carried out deep flank attacks through the jungle; and on Corregidor Island, the so-called Impregnable Rock lying in the entrance to Manila Bay, where the arrival of two Type 97s and a captured Stuart broke the back of the American resistance. But when they met opposition by their own kind it was a different story. The veteran 7th Armoured Brigade, consisting of 7th Hussars and 2 RTR, both equipped with Stuarts, reached Rangoon in February 1942 and thereafter covered the long retreat of I Burma Corps to India. Near Payagyi 7th Hussars destroyed a minimum of five Type 95s which, in the words of one officer, 'Appeared to be very lonely and untrained, and obviously did not know what to do, remaining stationary in the middle of an open field. They were knocked out immediately, before they knew we were there.' After this, the Japanese tanks either followed at a respectful distance or turned their attention to Lieutenant General Joseph Stilwell's Chinese divisions, which were withdrawing on a parallel axis.

More often than not, when they were specifically ordered to attack, the commanders of Japanese armoured units could think of nothing more imaginative than a mechanical version of the Banzai charge. One of the most spectacular of these took place on Saipan when the 9th Tank Regiment, with 44 Type 95s and Type 97s, counter-attacked the American beachhead at 0300 on 17 June 1944. Unfortunately for the attackers, they had been detected while forming up and their approach was illuminated by naval star shells. They ran into the combined fire of artillery, anti-tank guns, a Sherman platoon of the 2nd Marine Tank Battalion, and several M3 75mm tank destroyers. Inadequate reconnaissance resulted in several tanks wallowing into a nearby bog where they were picked off by bazooka teams. Only twelve Japanese tanks emerged from the engagement and all were destroyed shortly after.

Similar scenes took place on Luzon during the American liberation of the Philippine Islands. Here, although General Yamashita, the former conqueror of Malaya and Singapore, had the entire Japanese 2nd Armoured Division at his disposal, he so doubted its ability to engage in open warfare that he dispersed it in static groups which were ordered to impose delay on the enemy while, with the rest of his troops, he retired into the mountainous terrain in the north of the island. One such group, 45 tanks strong, was progressively pounded out of its emplacements at San Manuel until, during the early hours of 28 January 1945, its remaining 30 vehicles were destroyed as they mounted a suicidal attack, the cost to the Americans being three Shermans and one self-propelled howitzer. Between 1 and 7 February a larger group with 52 tanks was encircled at Munoz and shot to pieces by artillery and the Stuarts of 716th Tank Battalion when it attempted to break out with headlights blazing. At Lupao, some ten miles to the north, the 2nd Armoured Division's third major detachment, with 33 tanks, was destroyed within 24 hours in very similar circumstances; of the ten or eleven tanks that attempted to break out only five succeeded and these were later found abandoned in the foothills east of the town. Japanese sources later confirmed that by this time the 2nd Armoured Division had lost approximately 200 of its 220 tanks and was therefore finished as an armoured formation.

As with tanks, so with anti-tank guns. The Japanese Army had begun the war with an anti-tank rifle, for which it found little use, and a 37mm anti-tank gun that by 1941 was already obsolete, although of necessity it was to remain in service. Late that year the Model 01 47mm anti-tank gun also began to make its appearance and, as the Japanese series of runaway victories continued until the middle of 1942, no reservations were expressed about the weapon. Once Japan had been thrown onto the defensive, however, it became apparent that the Model 01, like the Type 97 medium tank, was lagging a generation behind its Western counterparts. At medium range it could penetrate the side armour of the Sherman and the Lee, while the Stuart was vulnerable everywhere; against the frontal armour of the Lee it produced results only at closer range, and for the frontal armour of the Sherman to be penetrated the range had to be closer still. There were, moreover, never enough Model 01s to go round. During the last months of the war the

Japanese caught up somewhat with their Model 90 75mm anti-tank gun, which could penetrate the Sherman at 1000 yards, but very few of these were built.

In general, Japanese anti-tank units had to make do with their 37mms and the few 47mms they were allocated, sometimes supplemented by field guns firing armour piercing ammunition, seconded from the divisional artillery. In the line, guns were expertly concealed and intelligently sited to cover a natural defile or bend in the road. Where no such features existed mines would be laid to channel enemy armour onto a tank killing ground; paddy fields were sometimes flooded for the same purpose.

Such situations were normal on any battlefield and could be dealt with by standard tactical drills. What most Allied crews found unsettling at first was the berserk fury, contempt for death and total dedication displayed by Japanese soldiers, either as individuals or in small groups, in attempts to destroy their tanks by various means at close quarters.

This was encountered almost as soon as British tanks arrived in Burma. While covering the withdrawal through Pegu, one of 7th Hussars' Stuarts was charged by a senior officer on a white horse. Sword in hand, the man scrambled aboard, but before he could do any damage the tank commander snatched a hammer from inside the turret and struck him a heavy blow on the head. The officer tumbled down the front of the moving vehicle and beneath one of the tracks, which passed over both his legs. Looking back, the commander saw him prop himself on one elbow, draw his pistol and fire several ineffective shots after the tank.

Later in the campaign, as the British Fourteenth Army was closing in on Mandalay, A Squadron 3rd Carabiniers was involved in a similar incident. The squadron was working in close, broken country in which one of its members recalled the Japanese were putting up 'the most fanatical defence against our tanks we ever encountered.' Suddenly a Japanese artillery lieutenant and a private burst out of the scrub and clambered aboard the rear of a Lee. The private was shot off by a burst of machine-gun fire from a neighbouring tank, but not before the officer had run the Lee's commander through with his sword. He then kicked the body down into the turret, dropped after it and killed the unsuspecting 37mm gunner in the same way. Only the 37mm gun breech separated him from the loader, Trooper Jenkins, who had just sufficient time to draw his revolver before the officer was slashing and stabbing at him. Jenkins sent all six bullets thudding into his opponent's body, but the man was quite berserk and refused to die. Jenkins closed with him and they fell on the bodies of the commander and gunner. At this point Jenkins noticed that the latter was lying with his pistol holster uppermost. He managed to free the weapon and fired three more shots at point blank range, finally settling the matter. Within the belly of the tank neither the 75mm crew, who continued to load and fire their weapon, nor the driver, had any idea of the carnage that was taking place behind them. Jenkins received the award of the Military Medal.

Well aware of their deficiencies in anti-tank armament, the Japanese formed what they called Human Combat Tank Destruction Squads, and in these the other ranks were just as capable of displaying suicidal courage as their officers. They attacked with pole charges, satchel charges and magnetic shaped charges, never with any chance of survival, hoping only that they would blow a tank to kingdom come. One such mass attack was mounted at the village of Oyin while Probyn's Horse, an Indian cavalry regiment equipped with Shermans, was converging on Meiktila with the rest of 255 Indian Tank

Brigade and 17th Indian Division in what proved to be the master stroke of the Burma campaign. The first tank hunter threw himself under a Sherman with his box of explosives, killing himself and the driver and disabling the vehicle. The second scrambled aboard another tank just too late to fling his box into the turret hatches, which were quickly slammed shut, and he was then shot off. The third threw himself under the same tank, which promptly reversed, leaving him curled round his charge in the road until it exploded. The fourth darted through a file of infantry and was climbing the glacis of another Sherman when his charge went off, blowing his body over the turret but causing no damage.

Of this incident the regiment's commanding officer, Lieutenant Colonel Miles Smeeton, wrote: 'I couldn't have been more surprised than I was at the sudden appearance of these Japanese soldiers, with their anguished look of determination and despair, pitting their puny strength against such tremendous force. This desperate form of courage was something that we knew little of and saw with amazement, admiration, and pity, too.'

As we have seen, there always had been plenty of Japanese willing to tackle a tank all but bare handed, long before these special tank hunting units, the Kamikazes of anti-tank warfare, were formed. For this reason it was quickly realised by all the Allies that whenever they were working in jungle or close country, tanks must have a close escort of infantry. In Burma, British and Indian tank units had the protection of the Bombay Grenadiers, whose battalions were specialists in the role. Often they were invisible to the tank crews, but the knowledge that they were always there was reassuring; some tank commanders willingly acknowledge that, on one occasion at least, they owed their lives to a Grenadier whose presence they never suspected. In street and village fighting, too, the Grenadiers were just as efficient in safeguarding their charges. For example, during the bitter battle for Meiktila in March 1944, the Japanese tank hunters concealed themselves in holes or craters in the road, a 100lb aerial bomb between their knees and a brick in one hand ready to strike the exposed detonator the minute a tank passed over; all were discovered and shot before they had the chance to do any damage. Some weeks later, while the Fourteenth Army was making its blitzkrieg style advance on Rangoon, the tank hunters adapted their tactics, burying the bombs nose upwards or using remotely detonated charges, and achieved a greater measure of success. In fierce fighting for the village of Payagale one Sherman of the Deccan Horse had its final drive assembly blown out, badly wounding the driver and hull gunner, and another had its engine blown through the decking and rolled over onto its turret roof. Yet, even in the latter case the infantry escort was close at hand, holding off the jubilant Japanese while the crew dug their way out of the turret hatches.

Nevertheless, no matter how conscientious the infantry escort might be, it was inevitable that, given their level of determination, some of the tank hunters would break through. Their most dangerous weapon was the magnetic shaped charge which, once fixed to the exterior of a tank, could blast its way through the thickest armour. In Burma this was countered by fitting stout wire grilles above the engine deck and glacis, so that the force of the explosion was dispersed before it reached the main armour, a form of defence also adopted by Matilda-equipped Australian armoured units in New Guinea, Bougainville and Borneo. The American-built Stuarts, Lees and Shermans could also be protected by laying spare links of their rubber-block tracks on the glacis, thereby reducing the area in which the magnetic attachment could take hold. By 1944 the experience of some US Marine Corps tank units in the Pacific had resulted in the hull sides of their Shermans being covered in oak planks, for just the same reason.

Hydrogen cyanide gas, contained in frangible glass grenades, was also used very briefly against tanks by the Japanese when they overran Burma. The idea was that when the grenade was smashed against a vision slit or hatch opening, the prevailing heat would ensure that some of the vapour would be drawn into the vehicle and incapacitate the crew. Obviously, some discretion had to be observed in the use of the weapon, not for humanitarian reasons, but because international opinion had banished gas from the battlefield, if not from top secret, closely guarded national stockpiles; Japanese reservations were therefore based solely on the fear that the Allies would discover its use and retaliate in kind, so creating a contingency which the Imperial Army was not adequately equipped to meet. However, the conquest of Burma was presenting problems which had not been encountered during the brilliantly successful campaign in Malaya; the Japanese armour had been intimidated, casualties among the infantry were proportionally higher and, within the Army, face was being lost. The cause of the trouble could clearly be identified as the efficient rearguard provided by 7th Armoured Brigade, and at some stage someone had banged the table and demanded that every possible means must be employed to eliminate its tanks. Someone else, obviously of very senior rank, had then authorised the limited use of the gas grenades.

On 14 and 16 April 1942 reports from two 7th Hussar patrol commanders mentioned that their tanks had been attacked at close quarters with 'glass phials.' In the first instance, involving Thai troops, the grenade had missed its target. In the second, a Japanese soldier jumped aboard the Stuart commanded by a Sergeant Campbell and managed to throw his grenade into the turret, where it broke. As the tank began to career out of control, Campbell, whose head had remained in the fresh air, glanced down and observed that the rest of his crew were unconscious. Dropping inside the turret, he slid forward into the driving compartment, in itself no easy task in the cramped interior of the Stuart, and, sitting on the driver, managed to manoeuvre the vehicle out of immediate danger and bring it to a standstill. He then returned to the turret and turned his guns on the ambush site. Campbell was subsequently awarded the Military Medal for his actions and his crew recovered. There is no record of gas grenades being used against tanks after this incident, very probably because the Japanese did not believe that the risk involved was justified by the results.

The concept of surrender was unintelligible to the Japanese soldier, who believed that to be taken alive by the enemy was the ultimate, unforgivable disgrace involving not only himself but his entire family. He therefore fought to the death as a matter of course, whether he was attacking over the bodies of his comrades or crouched behind his weapon in a bunker. His bunkers were masterpieces of field engineering, roofed with layers of logs and earth, often so well concealed that it was possible to stand on top of them without being aware of their existence. They were sited with interlocking arcs of fire and interconnected whenever possible, so that the defenders could move between fire positions without being detected. When time permitted those that were located on hills were provided with an approach tunnel dug through from the reverse slope. They could absorb any amount of punishment from field artillery without any apparent effect on the occupants. The only certain methods of neutralising them without incurring heavy infantry casualties were by a direct hit with a medium artillery shell, or by tanks employing close range gunfire. In the former case complete neutralisation of the entire bunker complex could not be guaranteed, while in the latter the approaches were generally mined and covered by concealed anti-tank guns.

Despite the obvious dangers, it was upon the tanks that the burden of bunker busting fell and it was to this task that a major part of their effort was directed in all the Pacific islands campaigns, in Papua/New Guinea, in the Arakan region of Burma, and in Manipur, India, where the Japanese sustained a major defeat in the linked battles of Imphal and Kohima. As Major General R. N. L. Hopkins commented in his history of the Royal Australian Armoured Corps, 'Without tanks the infantry very often found it impossible to close with the enemy. Where tanks could not be used, either because of bad ground or because there were none available, infantry casualties became crippling and the outcome was generally in doubt.' The words were an almost exact echo of Haig's when he described the results of the first tank attack on the Somme.

Some of the earliest actions of this type were those involving the Australian 2/6th Armoured Regiment in the fighting to eliminate the Japanese pocket at Buna on the northern coast of Papua. Here the going was so bad that the employment of tanks could only be justified by what the author of the Australian official history described as 'a desperate need.' Behind the beach lay a belt of swampland, cut up by tidal creeks and backed by thick jungle. Firmer ground existed in plantations and on two derelict airstrips, but even these were pitted by treacherous bog-holes and visibility was seriously restricted by head-high kunai grass. The latter also concealed the stumps of the trees which the Japanese had felled to construct their bunkers, and on these the tanks could belly and become as firmly stuck as if they had been impaled. Detailed route reconnaissance became as vital as it had been on World War I battlefields, but with drivers blinded by vegetation and tank commanders forced to work closed down because of sniper fire it was all too easy to miss landmarks and wallow off the chosen path.

A typical attack was made against enemy positions at Cape Endaiadere on 18 December 1942. Seven of the regiment's Stuarts took part, grinding their way slowly forward while they suppressed fire from bunkers and weapon slits. The Japanese responded with a Banzai charge, swamping several tanks under a wave of yelling infantry who fired into the vision slits, destroyed one with a magnetic charge and set fire to another after it bellied on a tree stump. They were shot off by the fire of the Australian infantry and the advance continued until the firmer going gave way to swamp.

Slow, attritional fighting of this kind continued until the middle of January 1943, when the last Japanese positions were overwhelmed. The cost in tanks and crews was high, much of the damage being caused by the point blank fire of concealed 37mm anti-tank and three-inch anti-aircraft guns. The performance of the Stuart's own 37mm main armament also left something to be desired. The HE round, having only slightly more force than a grenade, might enter a fire-slit and kill those closest to the explosion, but it caused little or no damage to the structure of a bunker; the AP round, on the other hand, could be used to splinter the timbers and bring down the roof over a fire-slit, but was of little use against personnel. The Australian conclusion that medium rather than light tanks were better suited to this type of close-quarter engagement coincided with that of the US Marine Corps, whose 1st Tank Battalion, also equipped with Stuarts, had simultaneously been fighting the same sort of battles on Guadalcanal, albeit on marginally better going. Having said that, it was readily accepted by both that the Stuart was the only tank which could have operated with any prospect of success in the prevailing conditions.

A year later a scientific method of bunker busting was developed in the Arakan by the 25th Dragoons. The British advance had been stalled near Razabil by a heavily fortified feature known as Tortoise Hill. Although the hill was bombed and shelled and some of its bunkers had apparently been knocked out by the Dragoons' Lees, infantry assaults met with an undiminished volume of return fire and failed to make much progress. At length it was discovered that as soon as the tanks and artillery opened fire the Japanese retired to the reverse slope of the hill; when the supporting fire lifted they rushed back into their bunkers to man automatic weapons and rain grenades on the advancing British and Indian riflemen. To counter this a drill was worked out. First, using HE, the tanks systematically blasted the scrub off the feature, exposing the bunker fire slits. Whenever possible, a smoke shell was posted through one of the slits and the smoke, seeping along the galleries, emerged from other, previously unsuspected, fire positions. The timbers of all visible slits were then smashed up with AP shot, causing their roofs to collapse. Simultaneously, the entire feature was being kept under sustained machine-gun fire and infantry's mortars were firing on the crest and reverse slopes, forcing the enemy to stay underground. Finally, the infantry stormed the slopes under a diversionary barrage of flat trajectory AP shot, which was maintained until the riflemen were within a few yards of their objective.

The 1945 campaign in the Arakan saw the drill being refined by the 19th Lancers, an Indian regiment equipped with the Sherman, which possessed more sophisticated gun control equipment than the Lee. A Forward Tank Officer (FTO) with a manpack radio was attached to the infantry, who could often see bunker slits that were invisible to the tank crews. By setting a datum point such as an easily identified tree, the FTO could adjust the tanks' fire using simple corrections, e.g., 'Right 100, drop 50,' until the target was hit, and then the drill would follow its usual course. The technique was subsequently employed against Chinese bunkers during the Korean War.

One of the major problems facing the Americans in the Pacific was that the islands were girdled by coral reefs through which landing craft had only limited access to the shoreline. To overcome this a family of amphibious vehicles was developed from a prototype designed by the American engineers Donald and John Roebling for use in rescue operations in the Florida Everglades. Known variously as Landing Vehicles Tracked (LVTs), amphibious tractors (amtracs), Buffaloes, Alligators or Fantails, their function was to swim as far as the reef, crawl across it, then negotiate the shallow lagoon beyond to deposit their troops on the shoreline. They undoubtedly saved the infantry many hundreds of casualties wherever they were employed yet, once they had landed, the infantry remained exposed to the murderous fire of the enemy's bunkers, and also vulnerable to counter-attack by Japanese armour, until their own tanks could be got ashore. A proportion of LVTs was therefore fitted with either the Stuart light tank turret or the turret of the Howitzer Motor Carriage M8, which mounted a short 75mm howitzer. Known as amtanks and formed into battalions, these vehicles were first committed to action in February 1944 during the landings in the Marshall Islands and thereafter took part in every major amphibious operation in the Pacific theatre of war. To preserve their buoyancy amtanks were thinly armoured, a fact not always appreciated by local commanders who insisted that they should follow the infantry inland, sometimes with unfortunate results.

As the fighting moved north towards the Japanese home islands the jungles were left behind and the terrain became more suitable for the mass deployment of armour. For example, on the tiny island of Iwo Jima, only eight miles square, no less than three Marine Corps tank battalions, equipped with a total of 200 Shermans including 27 flamethrowers, took part in the fighting. On Okinawa, the US Army and Marine Corps employed over 1000 amtracs and amtanks, 800 tanks and numerous self-propelled artillery weapons, APCs and tank destroyers. In both of these campaigns the Japanese integrated well-camouflaged natural caves into their defensive layout, often opening fire from them into the flank or rear of attacking infantry. This led to the tank crews developing another variation in bunker busting techniques, known as the Corkscrew and Blowtorch. First, the cave entrance was exposed by direct gunfire from tanks or tank destroyers. In itself, this might not prove sufficient to eliminate the defenders, who could simply retire deeper within the cave, so whenever possible a Sherman flamethrower was brought up to flame the interior, burning up the oxygen supply of those within even if they were beyond the reach of the flames. Finally, just in case there were survivors, the mouth of the cave was sealed with demolition charges.

Had it proved necessary for the Allies to invade Japan, the American contingent alone would have included three armoured divisions and at least 40 tank battalions, plus tank destroyer units and supporting armour of every type. Given the nature of Japanese resistance thus far, few expected the campaign to be anything but extremely costly. Despite the formidable expertise that they had acquired, it was fortunate that, for many of those who would have been involved, Japan surrendered in August 1945, following the atomic bomb attacks on Hiroshima and Nagasaki.

Above and Beyond...
Individual Achievements during the World Wars

'An army is a team,' said General George S. Patton. 'This individual heroic stuff is a lot of crap.' In the overall context, of course, the General, whose own nickname was Old Blood and Guts, spoke only the truth, although those who were familiar with the contrasting facets of his personality were not surprised when, awarding decorations to men of his Third Army who had distinguished themselves, he was sometimes inclined to display naked sentiment. In modern parlance, Patton might have said, in his own forthright style, that there is no place for a Rambo in an all-arms battlegroup; in the real world, Rambos are a liability who contribute very little and die quickly, often taking a number of their comrades with them.

Patton was speaking towards the end of World War II, during which all the combatant nations made full and frequent use of the exploits of air and submarine aces in their efforts to maintain morale. This was perfectly acceptable as the pilot and the submarine commander were, to a large extent, masters of their own destiny and were alone responsible for their achievements. In Germany, however, Dr Goebbels' propaganda ministry went a stage further and invented the tank ace, a concept which neither the British nor the American Armies believed to be valid. The problem was that tanks seldom fought in isolation and, more often than not, there were numerous influences at work around them which also had a bearing on the outcome of an action. Furthermore, because of breakdowns and other operational causes, a tank commander who was forced to change vehicles several times in the course of any one day would find himself working with different crews and, since manning a tank is in any event a team effort, it was obviously wrong that the entire credit for any collective achievement should be focused on one individual. Having said this, it can hardly be denied that men of outstanding courage and ability did exist, or that their deeds are worthy of record.

Such men were apparent from the beginnings of armoured warfare, one of the first examples being the crew of *F41 Fray Bentos*, a tank belonging to F Battalion Tank Corps that on 22 August 1917 took part in an attack during the Third Battle of Ypres. Together with three other tanks, *Fray Bentos*, named after a well-known brand of corned beef, made up No 9 Section, which was commanded by Captain Donald Richardson. The tank's regular crew consisted of 2nd Lieutenant George Hill, who preferred to drive himself and was good at it; Sergeant G. Missen, main gearsman; Privates P. Arthurs and J. Budd, six-pounder gunners; Privates W. Brady and J. Morrey, secondary gearsmen; and Private Trew and Lance Corporal Binley, Lewis gunners. Some were former infantrymen who had spent months in the line and knew exactly what to expect.

Some days before the attack, Richardson went to the front in order to study the ground over which his tanks would have to advance. It consisted of nothing but flooded shell craters with, here and there, the hulks of tanks lost in earlier attacks, sunk into the mire to the level of their sponson roofs. A sickly, all-pervading stench rose from the unburied bodies of men and horses, and bright sunlight was reflected from still water covered with green slime. Interviewed some forty years later by John Foley, Donald

Richardson could still recall the sense of great evil that hung over the apparently placid scene.

Richardson was aboard *Fray Bentos* when his section crossed the start line on 22 August. One of his tanks was immediately hit by the enemy's counter-barrage and began to burn fiercely. Gradually, *Fray Bentos* drew clear of the bursting shells. Ahead, the British bombardment was hammering the German trenches. Through the rear porthole Richardson could see infantrymen wading forward and, ominously, tanks beginning to wallow. Hill, however, was overcoming all obstacles. Soon, Richardson spotted his first objective, marked as Somme Farm on his map, and the tank slewed round towards it. At this point sustained machine-gun fire began rattling off the front of the tank, bullet splash penetrating the half closed visor to strike Hill in the hands and face. Budd, manning the port six-pounder, fired three rounds and the machine-gun stopped. Forty yards from the objective Richardson could see that the original buildings, long since pounded into rubble, had been replaced by a concrete pillbox approached by a communication trench. Budd opened fire again and four men ran out of the structure into the trench; using the forward Lewis gun, Richardson accounted for two of them.

He then decided that he would leave the infantry to mop up and continued towards his second objective, Gallipoli Farm. On an area of rising ground, however, traction was lost and progress slowed to a crawl. Richardson offered to relieve Hill, whose minor but extremely painful wounds had begun to swell. As they were changing places, machine-gun fire penetrated the visor, hitting both officers, Hill in the neck and Richardson in the thigh. In falling, Hill knocked open the throttle and the tank surged forward into a huge crater. Juddering, it almost succeeded in crossing the far lip but at the critical moment the engine stalled and it slid back. A further attempt to extricate the vehicle resulted only in *Fray Bentos* taking a steep list to starboard.

Richardson ordered Brady to clear the unditching beam from the roof. He left by the rear door and those within anxiously listened to him working overhead, knowing the risks involved. Their worst fears were realised when, following a burst of machine-gun fire from nearby, a thud was followed by the sound of something heavy slithering down the rear of the vehicle. Trew volunteered to make a second attempt, but Richardson would not allow it. Instead, he located the probable position of the enemy weapon and ordered Arthurs, the starboard six-pounder gunner, to engage it. Because of the list, this could only be done by fully depressing the breach. Arthurs fired twice before a shell exploded just beside his sponson. A steel splinter penetrated the gun port, slicing away his jaw and embedded itself in his chest. He obviously had little time left to live but the others did what they could for him. The artillery fire intensified and it was soon apparent that the crater was proving to be the tank's salvation, for the enemy's shells were exploding just short of or just beyond the vehicle.

By early afternoon the list had become so pronounced that the starboard six-pounder was pointing into the mud and its port counterpart skywards. The shelling stopped and it became apparent that the British attack had petered out. Richardson and Budd decided to try and attach the unditching beam, but were unable to open the rear door more than a few inches because Brady's body and the massive beam were lying against it.

It was decided to try and move the tank again under its own power. This proved to be disastrous, for the earth suddenly gave way beneath the starboard track, causing the muzzle of the six-pounder to strike the ground so that the heavy steel breach swung without warning. It struck Budd a tremendous blow on the chest, shattering his rib cage, and blood began to flow from his mouth, indicating severe internal injuries.

Trew, manning the forward Lewis gun, suddenly opened fire on a German counter-attack force gathering near Gallipoli Farm, breaking it up. After that the shelling recommenced. Conditions within *Fray Bentos* began to deteriorate. The summer sun turned the interior into an oven filled with combined smell of machinery and burned cordite; slime was flooding through the starboard sponson, mingling with blood and leaking oil; flies penetrated the smallest apertures to batten on wounds. Hill, though clearly in a bad way, had recovered consciousness and was able to swallow a little rum and water while the others broke open the ration box and fed themselves on corned beef, hard-tack biscuits and quince jam. At about 1800 the German artillery, having wasted much ammunition without being able to close the bracket, ceased firing; its place was taken by a trench mortar which, although it was the ideal weapon in the circumstances, still failed to score anything better than a near miss. Shortly before dusk Morrey, now taking his turn on the Lewis gun, broke up another counter-attack force near the farm.

Small arms fire had begun striking the rear of the tank. Richardson was sure that it came from his own lines and he knew that, following standard procedure, the British artillery would destroy the apparent derelict to prevent it falling into enemy hands. Sergeant Missen volunteered to go back and inform them that the tank was still manned, clambering through the port sponson door as soon as it was dark; though badly wounded on the way, he managed to get through.

Richardson had no intention of abandoning the vehicle as long as he could cause the enemy some trouble, and he also harboured a faint hope that somehow a tow could be got up to them. For their part, the Germans clearly regarded *Fray Bentos* as a thorn in their side, and shortly after 0100 they despatched a party to wreck the vehicle. They were no sooner detected by Morrey, who opened fire with the Lewis gun, than the port sponson door was wrenched open. Richardson had a brief impression of figure in a coal-scuttle helmet, a stick grenade in his hand. He fired one shot with his revolver, saw the figure topple from sight, and then Trew clamped the door shut. A demolition charge exploded beneath the sponson, shearing the holding bolts and pushing it inwards along its rail transportation runners, striking Trew a heavy blow on the head and shoulders. Binley had slammed the rear door as soon the raid began, but Richardson made him reopen it wide enough for him to use his revolver. Two men appeared, carrying a box which he took to be another demolition charge; he fired five shots and they ran off, dropping it. There were no further attacks, but the tank continued to settle until the oozing mud covered the entire floor. Budd died shortly before dawn.

To preserve vision, the port sponson door was propped open with a shell case and the survivors took turns manning the aperture. While so doing Binley had part of his scalp removed by a shell splinter. As the supply of drinking water was all but exhausted Richardson opened the radiator drain cock from time to time; the orange coloured liquid left a foul aftertaste, but it was better than nothing. Somehow, Richardson managed to preserve his men's morale by persuading them that they were hurting the enemy. As if to emphasise the point, during the afternoon a third German counter-attack force was seen forming up some way to the left of the vehicle. The Lewis gun would not bear, so the crew sniped at it with a rifle through the crack in the sponson door and the group scattered. In response, *Fray Bentos* was subjected to further machine-gunning and mortaring.

At about 0100 that night Morrey, positioned at the rear door, reported that he could see figures moving towards the rear of the tank from the direction of the British front line. For a while Richardson was uncertain whether they were British or German. The sudden bursting of a green flare overhead revealed that they were the latter. He fired at them with his revolver and Morrey joined in with a rifle but the enemy had closed in to within twenty yards when, quite unexpectedly, several machine-guns opened fire from the British line, cutting some of them down and scattering the rest. Thereafter, flares were put up at regular intervals to illuminate the tank and the attack was not renewed. It was reassuring to receive such confirmation that they had not been forgotten.

They passed the third day in a semi-comatose state induced by lack of sleep, the effect of untreated wounds and the hot, fetid atmosphere within the vehicle. The mud climbed higher and any attempt to open the sponson door immediately attracted more mortar and machine-gun fire. At one point a mortar round struck the left hand track, Trew's face being slashed open by a flying sliver of metal. The effect of the explosion was deafening and the interior was filled with choking fumes. When it cleared a patch of sky could be seen through a jagged hole in the plating. The mortar gunner, however, was evidently unaware that he had at last found the range, for his success was not repeated.

Richardson discussed the situation with Hill, who was now much stronger. They had no rations, there were only a few rounds of rifle and revolver ammunition left, and not much more for the Lewis gun. Of the entire crew, only Morrey remained unwounded. They had fought the tank to the limit and in the process caused the local German commander some grief. It was decided, therefore that they would abandon the vehicle after dark. To everyone's surprise, when they did so, taking the unshipped Lewis guns with them, the enemy made no attempt to interfere with them. Richardson left last, experiencing a curious sense of loss as he looked round for the last time. Hill had run a tight ship and only 72 hours earlier the interior of the tank had gleamed like the engine room of a small warship; now, flooded with slime and splattered with blood, *Fray Bentos* was to be left with only the dead for a crew. He crossed no-man's-land at the height of the evening hate, hobbling into the infantry lines to be told that the others had also reached safety. Three months later, his wound healed, he fought in the great tank attack at Cambrai.

One of the most famous stories in the history of the Tank Corps is that of Lieutenant Clement Arnold and his Whippet tank *Musical Box* during the first day of the Battle of Amiens. Arnold's family ran a well-known drapery business in Llandudno, North Wales, and as a member of the original D Company his brother Arthur had taken part in the world's first tank attack, described in an earlier chapter, winning the Military Cross.

Musical Box formed part of the 6th Battalion's B Company, the task of which was to support the cavalry as it exploited the breakthrough. By 0620 the tank was 2000 yards ahead of the original start line and was beginning to catch up with the heavy tanks and the Australian infantry. Between Abancourt and Bayonvillers it came under fire from a four-gun battery that had already knocked out two Mark Vs nearby. Using a belt of trees for cover Arnold swept round behind the battery. Seeing this, the gunners attempted to flee but were cut down by Arnold and his gunner, Private Ribbans. The Australians arrived to occupy the position and Arnold set off again, travelling east to the north of and parallel with the Amiens-Paris railway line. By now the cavalry had come forward and were beginning to harry fugitives from the broken enemy lines. Unfortunately, whenever they met stiffer opposition the horsemen began to suffer casualties and were forced to halt. Arnold provided assistance for two such patrols, killing most of their opponents, and the troopers cantered on towards their next check.

At this point Arnold evidently decided that he would make far better progress on his own. His map told that there was an enemy bivouac area in a shallow valley between Bayonvillers and Harbonnières. He drove into this to find the occupants hastily stuffing their belongings into their packs and opened fire at once, killing or disabling about 60 of them before the rest made good their escape across the railway embankment.

At this point the railway swung away to the south-east and Arnold continued due east across country. He soon came across files of infantry retreating through the standing corn and engaged them at ranges of between 200 and 600 yards. *Musical Box* remained in the area for over an hour, shooting at any sign of movement as it cruised up and down. In the process, return fire punctured the spare cans of petrol that were stowed on the vehicle's roof. The contents flowed down the inside walls of the fighting compartment, rendering the air, already polluted by the incessant firing of the machine-guns, so foul that the crew were forced to don their gas masks.

Because of the risk of fire and explosion most tank commanders would have decided that enough was enough. Arnold, however, had the bit between his teeth and, although he was now completely alone, he decided to press on. Soon he found himself in the midst of the retreating army, surrounded by columns of horse and motor transport and marching men who believed that they were beyond the reach of danger. When *Musical Box* opened fire on them at close quarters, inflicting heavy losses, a wild panic ensued.

Now satisfied, Arnold ordered his driver, Private Carney, to turn for home, but he had left it a moment too late. The enemy had brought up a field gun which quickly scored three hits, causing the tank to burst into flames. The crew leapt clear and rolled on the ground amid pools of blazing petrol in a desperate attempt to extinguish their burning clothes. Enraged, the Germans swarmed round, shooting Carney dead while Arnold and Ribbans were kicked and hammered with rifle butts until they were all but senseless. Arnold's refusal to answer his interrogator's questions earned him five days in solitary confinement on minimum rations, and he was then shipped to Freiburg prisoner of war camp where he met his brother Arthur.

The burned-out shell of *Musical Box* was found by the advancing Australians some eight miles beyond what had been the original start line. Only on Arnold's release when the war ended three months later did the full story become known. He was awarded the DSO but would also enter the record books as the first officer to carry out an exploitation in a tracked fighting vehicle, the results being comparable to those achieved by the armoured cars of the 17th Battalion elsewhere on the same battlefield, which have also been recorded in a previous chapter.

Rather similar to Arnold's exploit was that of Captain Darryl Hollands during the fight at Steamroller Farm near El Aroussa, Tunisia, on 28 February 1943. The previous evening a German counter-offensive had cut the road between El Aroussa and Med-jez el Bab. The commander of Y Division, a scratch force holding this sector of the front, had no idea of the enemy's strength and he decided to probe the area of Steamroller Farm, four and a half miles to the north, and the pass lying immediately beyond. The troops detailed for the task included a company of the 2nd Coldstream Guards, A Squadron 51 RTR, equipped with Churchills, and a troop of field guns. Hollands, a quiet man with immense resources of physical courage, had already been awarded the Distinguished Conduct Medal when, as a sergeant, he had rescued a pilot from a burning aircraft at Dunkirk; now, he was commanding A Squadron's 1 Troop.

As the force approached Steamroller Farm it was apparent that it was heavily defended and the squadron came under fire from an anti-tank gun screen sited beyond an intervening wadi. A fierce fire fight ensued, during which the tanks were also dive-bombed by a Stuka squadron. At this point the squadron commander, Major E. W. H. Hadfield, was ordered to break through 'at all costs' and secure the high ground around the pass. Hadfield now had only nine of his Churchills left and he felt that these would be sacrificed to no purpose if they were committed to such an attack. Neverthe-less, he had to comply in some way and ordered 1 Troop to advance on the objective, for-getting that casualties had reduced its strength to a single tank, Hollands' own *Adventurer*.

The chances of survival, let alone success, were very slim, and in these circum-stances Hollands' reaction was to give himself up for dead, get on with the job, and do as much damage as he could in the process. *Adventurer* moved forward, only to find the way ahead blocked by the wadi. Hollands told his driver, Trooper John Mitton, to reverse some little distance, and then swung the tank towards the road on the right, which crossed the wadi on a causeway. This placed the vehicle broadside on to the enemy, but fortunately the latter's view was interrupted by scrub and their shots cracked harmlessly past. Reach-ing the road, *Adventurer* turned left, heading for the causeway, and had just rounded a bend where the scrub ended when it came face-to-face with an Eighty-Eight at thirty yards' range. The tank rocked to a standstill and got in the first shot, wrecking the gun.

Hollands set off again, believing that he had had his share of luck for the day. *Adventurer* roared across the causeway, turret traversed left towards the enemy. The road began to climb and, rounding a double bend, Mitton braked sharply when he encoun-tered a barrier of camouflage. For the second time in minutes he was confronted by the yawning black muzzle of an Eighty-Eight. It suddenly vanished within a huge belch of flame. From within the vehicle came a clatter of falling equipment. The round had sim-ply scoured its way along the top of the turret, tearing away the extractor fan casing and smashing the rear stowage bin. In the turret the gunner was frantically struggling to bring the main armament to bear but was thwarted by a loose round that was temporarily foul-ing the traverse gear. The black muzzle belched flame again, but for some reason the sec-ond round missed completely. Meanwhile, Trooper Hank Howsen, the hull gunner, was methodically loading a fresh belt into his Besa machine-gun. He snapped the cover shut, pulled back the cocking handle, laid the weapon and fired a long burst through the Eighty-Eight's camouflage. The gun crew took to their heels pursued by *Adventurer* with Hollands firing his Thompson sub-machine-gun and hurling grenades at them from the turret.

Having twice survived sudden death by a whisker, Hollands was fighting mad. Clearing his jammed traverse gear, he turned left off the road and headed for some infantry positions on high ground near pine trees, overrunning slit trenches on the way to the crest. From this, he could see that the enemy had parked their transport, amounting to 27 vehicles, further along the road behind a spur, and his gunners set them ablaze. As *Adventurer* was now completely alone in the heart of the German position he asked Hadfield to reinforce him as quickly as possible. However, so intent was Hollands on his work of destruction that he was barely aware of the passage of enemy aircraft overhead at about 1730. They dropped paratroops to reinforce the garrison of Steamroller Farm, with the result that A Squadron found itself even busier than before. Hadfield could spare only one tank, that of Lieutenant J. G. Renton, who set off along the same route.

Meanwhile, Hollands had become involved in a very personal duel. A German in a camouflaged slit trench immediately ahead of *Adventurer* kept bobbing up and shooting at the tank with a rifle grenade thrower. Two belts of Besa and three armour piercing rounds failed to solve the problem. Mitton recalls that when a fourth AP was fired into the ground just short of the trench, 'It seemed to vanish in a cloud of smoke and dust. The net flapped wildly, breaking free from the blast. When the dust settled the German crawled out, stood looking dazedly at the tank, turned slowly, dropped his rifle and staggered away.' No one aboard was inclined to take the matter further.

At this point two PzKw IIIs appeared close to the head of the pass. Hollands was unable to depress his main armament sufficiently to engage them but Renton had now come up alongside and his gunner, Trooper Nicholson, put three rounds into each. Shortly after 1800 Hollands received the order to withdraw. Hardly had the move begun than his radio failed completely and with it the intercom. Climbing out of the turret, he sat on the front of the vehicle, directing Mitton with hand signals through the open visor. When they stopped briefly to put two AP rounds into the second Eighty-Eight, the engine stalled and refused to restart. Renton overtook and the two crews attached a tow chain under mortar and machine-gun fire, Renton being wounded as he scrambled back aboard. *Adventurer's* engine started at the first pull and the two tanks succeeded in reaching their own lines, picking up the crew of a burning Churchill on the way.

As the combined force was clearly too small to capture the objective, it was ordered to withdraw by the commander of Y Division. A Squadron's losses amounted to three killed, eleven wounded, three tanks destroyed, two disabled and the rest damaged to a greater or lesser degree. At first, no one was inclined to believe Hollands' and Renton's story, but three days later the infantry took possession of the area and reported even greater damage than had been claimed, including two PzKw IIIs, eight anti-tank guns, two light anti-aircraft guns, two mortars, 25 assorted vehicles and up to 200 personnel casualties. The Fight at Steamroller Farm had effectively destroyed the enemy's chances of taking El Aroussa; an intercepted radio message from the German battlegroup commander, who was evidently unfamiliar with the characteristics of the Churchill, justified his withdrawal on the grounds that he had been attacked by 'a mad tank battalion that had scaled impossible heights.' Hollands received the Distinguished Service Order, Renton the Military Cross and Mitton the Military Medal.

Asked to name a German tank ace, most Western students of World War II would probably suggest SS Lieutenant Michael Wittmann because of the spectacular manner in which he halted the drive of the British 7th Armoured Division near Villers Bocage, Normandy, on 13 June 1944. A regular soldier, Wittmann had joined the Army in 1934 and transferred to the Waffen SS two years later. He served in armoured cars and assault guns and was commissioned on 12 December 1942. The following month he began his long association with the Tiger E, serving with 13 Company of SS Panzer Regiment 1 Leibstandarte Adolf Hitler, with which he fought at Kursk. In February 1944 he assumed command of 2 Company SS Heavy Tank Battalion 101, part of I SS Panzer Corps. By the time his unit moved to Normandy he had been credited with the destruction of 119 Russian tanks and had been awarded the Iron Cross 1st and 2nd Class and the Knight's Cross with Oak Leaves.

On the morning of 13 June Wittmann's company, numbering six Tigers of which only four were in a condition to fight, was positioned on the reverse slopes of Hill 213, overlooking Villers Bocage. Shortly after dawn the leading brigade of 7th Armoured Division, engaged in a wide flanking move intended to isolate the city of Caen, approached the village. The advance guard, consisting of A Squadron 4th County of London Yeomanry and A Company 1st Rifle Brigade, passed through and halted on the road beyond. The tanks, carriers and half-tracks were then pulled in to the side and closed up nose-to-tail to allow room for the relief point units to pass through. Given the nature of the country and the known proximity of the enemy, this was not the wisest course of action, since it effectively denied movement to most of the vehicles in the column.

Wittmann, watching the entire process from the summit of the hill, recognised an opportunity that was too good to miss. His Tiger lumbered out of cover onto a track that ran parallel to the road and his first round blasted a half-track, blazing fiercely, across the highway, blocking it. He then travelled slowly along the length of the column, his gun claiming a further victim with each round it fired. Return fire simply bounced off the thick armour and, gaining the road, he ground his way into the village, destroying the Yeomanry's RHQ Troop and several artillery observation Shermans. A Cromwell reversed into a side road, hoping to put a shot through the Tiger's thinner rear armour, but Wittmann, anticipating the move, had his turret traversed ready and got in the first fatal shot. He then turned off the main street and returned to Hill 213 across country.

The afternoon's fighting was not quite so successful. 2 Company, supported by elements of 2nd Panzer Division, broke into Villers Bocage but in fierce street fighting had three Tigers knocked out and a further three, including Wittmann's, immobilised. Nevertheless, the 7th Armoured Division's had lost 25 tanks, 14 half-tracks and 14 carriers, the majority destroyed during the morning, and after dusk it withdrew to the high ground west of the village.

As a result of the episode, Wittmann was awarded the Swords to his Knight's Cross and promoted to captain. He declined an appointment to an officers' tactical school and was killed on 8 August while engaged in a fight with a troop of British Sherman Fireflies; the generally accepted version is that his Tiger fell victim to the seventeen-pounder guns of the latter, but one German source suggests that it was destroyed either by a heavy calibre artillery round or a rocket fired by a Typhoon ground-attack aircraft.

In contrast to Wittmann, comparatively few in the West have ever heard of Count Hyazinth Strachwitz von Gross-Zauche und Camminetz, for the simple reason that from 1941 until 1945 he fought solely on the Russian Front. Even had the ace system not existed, Count Strachwitz possessed a natural mastery of the tactical battle that ensured him a prominent place among the German armoured corps' regimental commanders. In succession, he was awarded every grade of the Iron Cross up to and including the Oak Leaves, Swords and Diamonds of the Knight's Cross, attributing his numerous successes to a sixth sense, which obviously included a clear understanding of the Russian mind.

Born in 1893, he belonged to an old Silesian noble family, no fewer than ten members of which had died fighting the Mongols at Liegnitz in 1241. His unusual Christian name had been conferred on the family's first-born for the past 700 years in honour of the Silesian saint to whom the chapel in their castle at Grossstein was dedicated. He followed the traditional route into the Imperial Army, being sponsored personally by Kaiser Wilhelm II for a commission in the élite Garde du Corps cuirassier regiment and was educated at the famous Lichterfelde Cadet Academy where one of his friends was Freiherr Manfred von Richthofen, the future fighter ace. An outstanding rider, swordsman and athlete, he would have been a candidate for the German Olympic team of 1916, had not the war intervened. In 1914, after leading a week-long foray through the French rear areas, he was eventually surrounded and forced to surrender, remaining a prisoner for the rest of the war. Returning home in 1918 he found Germany changed beyond recognition and fought with the Freikorps to restore stability to Silesia, parts of which were claimed by the newly created state of Poland. He retained a reserve commission in the Reichsheer's Cavalry Regiment 7, the eight squadrons of which each maintained the traditions of one of the former regiments of the Imperial Army. In 1935 he joined a large cadre that was transferred to the newly created Panzer Regiment 2, with which he fought in the Polish, French and Balkan campaigns, often demonstrating a startling tactical originality.

It was, however, as commander of the regiment's 1st Battalion in Russia that Strachwitz began to attract the public attention. Courteous and urbane, he nevertheless drove his men hard, recognising that movement often provided its own defence, even if the cost was continuous lack of sleep and rest. 'Tanks must not be allowed to stand still,' was his dictum, 'They must be permanently on the move and always led from the front.' This he continued to do throughout the war. In October 1944, after resolving a difficult situation near Riga, he halted his tank beside three senior officers.

'Bravo, Lieutenant, well done!' said one, as he threw back his cupola hatch to reveal an oil-streaked face.

'You're not talking to a lieutenant,' replied Strachwitz equably, 'I'm only a general!'

To his colleagues he was known simply as the Count, but the press took to referring to him as Der Panzergraf, just as, a generation earlier, they had, with slight inaccuracy, named his old friend Richthofen The Red Baron.

His part in the great tank battle at Brody-Dubno has been recounted elsewhere, but this was just the first of many episodes of which it is possible to record only a few. In fact, some days prior to this action his tank had broken down and Russian infantry closed in around it. Strachwitz took the initiative and, dismounting, chased them off while his crew completed the repair, incurring a wound in the process.

Driving his battalion on by day and night, his advance often outstripped the Russian withdrawal. He impressed upon his officers the need to lull the Russians into a false sense of security until the last possible moment. Sometimes his tanks travelled with their guns deliberately traversed to the rear and the Russians, believing them to be their own, allowed them to penetrate their defences; by the time the truth was discovered it was too late. When the Bug was bridged he led his battalion in a headlong dash into the enemy's rear, running down and destroying a column of 300 vehicles, including several anti-tank and artillery units that believed themselves to be on the road to safety.

During the 1942 campaign Strachwitz, now a lieutenant colonel, reached Stalingrad with the vanguard of the Sixth Army. Luckily, perhaps, he was seriously wounded in the heavy fighting for the city and was evacuated, thus avoiding the fate of his comrades in the subsequent débâcle. On returning to duty he was again promoted and appointed commander of the élite Grossdeutschland Division's armoured regiment, the 3rd Battalion which was equipped with Tigers. After the failure of the Kursk offensive his regiment took part in the ferocious battles around Kharkov, one of which has already been described. Strachwitz now proved himself to be as formidable a tactician in defence as he had been in attack. He noticed that in executing their rigidly controlled attacks, Soviet tank commanders tended to look straight ahead, rarely to the side and almost never behind; this ensured the success of counter-attacks into the attackers' flanks and rear and also enabled the defenders to prepare tank killing grounds. Strachwitz was particularly adept at the latter and on one occasion predicted with uncanny accuracy the line of advance the Russians would take in a night attack, placing his Tigers in concealed positions with strict orders neither to move nor fire until he did so himself. During the early hours of the morning eighty or so T34s trundled onto the killing ground, convinced by the silence that the Germans had abandoned the village that was their objective. Strachwitz let his own target close in to within 40 yards before blowing its turret off with his first round. Within minutes 36 of the Soviet tanks had been reduced to burning wrecks and the rest had vanished into the darkness.

On 1 April 1944 he was promoted to major general, a hitherto unheard-of elevation for a reserve officer. He commanded the 1st Panzer Division briefly and in the autumn was appointed Commander (Armour) to Army Group North. This involved an almost continuous process of gathering together such armoured and Panzergrenadier units as he was able in order to shore up one sector of the crumbling front after another. 'Strachwitz is here – he'll sort it out!' was the reaction of more than one relieved local commander when his battlegroup arrived.

It was during this period that the Count achieved one of his greatest coups. The front was temporarily quiet and once again he predicted the route along which the Russians would renew their offensive. With only four tanks he penetrated deep into their rear and carefully concealed his vehicles around the chosen killing ground. In due course the first enemy tank units appeared, rolling towards the front, completely oblivious of the German presence, and were quickly destroyed. More followed and met the same fate. Unable to identify the source of their destruction, each Russian wave continued to believe that it lay to their front, when in fact it lay behind them. Incredibly, the Soviet commanders permitted the inexplicable massacre to continue; an hour after the first shot had been fired 105 of their tanks lay burning on the killing ground. Equally incredible was the fact that Strachwitz was able to withdraw his tiny unit intact.

Yet, not even a Strachwitz could hold back the Russian tide forever. As the war drew towards its inevitable end he dismissed what remained of his command, telling the men to surrender to the Western Allies. At the personal level, the war had brought him nothing but tragedy. He had lost both his sons and while he was a prisoner in American hands his wife was killed in a traffic accident; his home now lay far to the east, lost for-ever. Gradually he rebuilt his life, married again and was blessed with a second family. Latterly he lived on a small estate at Chiemsee, a pleasant part of Bavaria where the eccen-tric Ludwig II had built one of his fairytale palaces on an island in the lake. He was respected not only for his achievements, but also as an officer of the old school who remained above politics, putting his duty to his country first and foremost. It was in recognition of this that, when he died in 1968, officers of the Bundeswehr mounted guard around his coffin.

Red Sheet
The Sinai Campaign of 1967

s a small nation surrounded by enemies, Israel was acutely conscious that if she was to survive she could not afford to lose a battle, much less a war. Of necessity, therefore, the Israeli Defence Force (IDF) was a citizen army in which all of military age were liable to serve, accepting compulsory periods of regular service, reserve training and periodic mobilisations as a normal part of their lives. Equally, it was necessary for Israel to be armed to the teeth, most of her weapons being supplied by the United States, the United Kingdom and France. The organisation of the IDF's tactical formations followed the flexible patterns of the US Army, while in action the use of 'saddle orders' and ad hoc battlegroups in the German manner was well suited to the Israeli temperament. The result was that within any overall plan of campaign the IDF, highly motivated and capable of thinking and acting on the move, possessed the capacity to exploit local situations to its own advantage.

Israel's Arab enemies were also highly motivated, although for them the loss of a battle or a war did not threaten their survival. After the 1956 war, in which the UK and France had colluded with Israel to launch an attack on Egypt, it was natural that many of the Arab nations should turn to the Soviet Union for arms and military assistance, both of which were provided on generous terms. The Soviet tanks of the period, notably the T54/55 series, were less sophisticated than their Western counterparts, but they were far less expensive and their simplicity made the training of Arab crews, many of whom lacked the technical education of the Israelis, a relatively quick and simple matter. Because of these factors, most Arab general staffs were prepared to overlook the principal disadvantage of the T54/55, which was a low, domed turret that seriously restricted the upwards movement of the main armament breech, thus curtailing the degree of depression that could be obtained and so reducing the tank's ability to fight hull-down. With Soviet equipment came Soviet command methods that emphasised the importance of central control and restricted the degree of personal initiative allowed to commanders on the spot, the consequence being that operations were conducted in accordance with a preconceived plan at a slower tempo than that of the IDF. Needless to say, if the overall plan was disjointed by enemy action, the stiffness inherent in the system often prevented remedial action being taken before the situation degenerated until it was beyond control. A notable exception was the small but efficient Royal Jordanian Army, which had preserved many of the traditions of the old British-officered Arab Legion. The Jordanians, among whom the British influence remained strong, preferred to equip their armoured units with Centurions and Pattons and on occasion came very close to inflicting a defeat on the IDF.

It was, however, in Sinai that the Israelis and Egyptians fought the greatest tank battles since World War II, those of 1967 demonstrating most clearly the strengths and weaknesses of the two sides. The Sinai peninsula is triangular in shape, having a maximum width of 130 miles along its Mediterranean coast and a length of 240 miles from north to south. In the east it is separated from Saudi Arabia by the Gulf of Aqaba, and in the west it is divided from Egypt proper by the Gulf of Suez and the Suez Canal. From the northern shore the ground rises steadily across a sand, gravel and rock plateau to the 8664 foot peak of the Mount Sinai massif, then falls steeply towards the Red Sea. Three routes cross the peninsula between the Israeli/Egyptian frontier and the Canal. Of these the most northerly is the shortest and best, following the coast from Gaza through Rafah, El Arish and Romani to El Kantara. South of and roughly parallel to this a road runs from Kusseima to Abu Agheila and on through Bir Gifgafa and the Tassa Pass to Ismailia. Further south still, a more difficult track exists between El Kuntilla, Thamad and Nakhl, then winds on through the Mitla Pass to the town of Suez at the southern end of the Canal. North-south communication is restricted to a track running from El Arish through Bir Lahfan to Abu Agheila and on past Djebel Libni to Nakhl, with a branch diverging in a south-westerly direction from Abu Agheila to Bir Hasana and the Mitla Pass. The landscape is everywhere hot, parched and sterile, and since it absorbs less than ten inches of seasonal rainfall each year sources of water are few and far between.

Since 1956 Gamal Abdel Nasser, the President of Egypt, had been the hero and leader of the more radical elements within the Arab world. His claim that the Israeli victory in Sinai that year would have been impossible without British and French support was widely believed and since then, with Soviet assistance, he had made Egypt a major military power. Likewise, his determination that Israel would be destroyed and dismem-

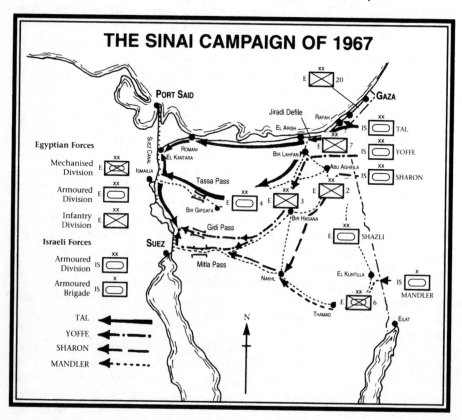

THE SINAI CAMPAIGN OF 1967

bered among the victorious Arabs, broadcast with increasing stridency by Radio Cairo, won him enormous support throughout the Middle East. Assured of the efficiency of his newly equipped army by his Soviet advisers, he became the victim of his own propaganda and by May 1967 had deployed no less than 100,000 in Sinai, so creating a threat to which Israel was bound to react.

General Abdul Mortagy, the Egyptian Commander in Chief in Sinai, had studied several of Montgomery's set piece battles, notably Alam Halfa, and he was also known to favour the Soviet Army's concept of a defence in depth intended to wear down an attacker's strength, followed by an armoured riposte that would inflict a decisive defeat, very much in the manner of Kursk. He predicted, correctly, that the Israelis would strike the first blow and there is every reason to believe that, left to himself, he would have established his defensive belt in western Sinai. The effect of this would, partially at least, have offset the stiffness in his own command structure and it would have ensured not only that the IDF would have to open its attack at the end of a very long line of communications, but also that in the event of failure it would have to conduct a difficult withdrawal across many miles of desert.

However sensible they might have been, such ideas were not acceptable to Nasser, since they involved the apparent abandonment of large areas of Egyptian territory before the first shot had been fired. The fact that these areas possessed no military value, and that the defence of eastern Sinai would leave the Egyptians with open desert at their back, was considered to be less important than projecting the image of an army poised on the brink of victory. Mortagy was forced, therefore, to make the same sort of forward deployment that had failed in 1956, albeit in greater strength and depth. The 20th (Palestinian) Division, with 50 Shermans, was holding the Gaza Strip; the 7th Infantry Division, with 100 T34/85s and JS IIIs, was responsible for the defence of Rafah, the Jiradi Defile, where the coast road passed through an area of apparently impassable dunes, and El Arish; the important track junction at Abu Agheila was held by the 2nd Infantry Division with the 3rd Infantry Division deployed in depth to the west near Djebel Libni, each with 100 T34/85s and T54s; to the south the 6th Mechanised Division, also with 100 T34/85s and T54s, covered the axis El Kuntilla-Nakhl; at Bir Gifgafa, roughly in the centre of the peninsula, was Major General Sidki el Ghoul's 4th Armoured Division, ready to deliver the armoured counter-stroke with its 200 T55s; and between Kusseima and El Kuntilla was a second armoured formation, named Task Force Shazli after its commander, Major General Saad el Din Shazli, equipped with 150 T55s, which was to cross the Israeli frontier into the Negev Desert and isolate the port of Eilat.

With the exception of Task Force Shazli, this deployment was primarily defensive in character and tacitly surrendered the initiative to the IDF. Again, although numerous historical precedents emphasised the importance of the operative level of command in desert warfare, Mortagy alone was responsible for coordinating the operations of seven formations from his headquarters far to the rear, a task which could have been eased considerably if he had established an intermediate corps headquarters with a degree of local autonomy. Furthermore, while the Egyptians had 800 tanks in the line, with a further 150 in reserve, only 350 of these were serving with armoured formations while the rest were subordinate to local infantry commanders who often reduced their potential contribution by digging them into static defence systems. Finally, the assumption that the

Israelis would willingly engage in a contest of attrition was fundamentally flawed; it would have been safer to assume that once the initiative had been surrendered to the IDF, which was always conscious of its limited manpower resources, the Israelis would impose their own conditions on the battle.

The Israeli Armoured Corps had won its spurs in the 1956 war and since then it had been regarded by the IDF as the decisive arm in the land battle. It had been expanded and was now equipped with Centurions and M48 Pattons, all of the former and many of the latter being upgunned with the excellent British 105mm tank gun, as well as upgunned Shermans and French AMX-13s. The IDF had begun mobilising its reserves on 20 May and by the time this process was complete Major General Yeshayahu Gavish's Southern Command had three armoured divisions plus two small independent armoured brigades in the line opposite the Egyptians. This produced a total tank strength of 680 with 70 in immediate reserve, with all but a few serving within armoured formations. Gavish's orders were breathtakingly simple in their scope – his armour was to smash through the enemy's defences on parallel axes and advance rapidly to the Suez Canal, which would become Israel's new and defensible military frontier with Egypt; the Egyptian Army, fragmented and cut off from its homeland, would wither and die in the desert.

On the right of the Israeli line, opposite the Gaza Strip and Rafah, was Major General Israel Tal's armoured division. Tal had served as the platoon sergeant of a machine-gun platoon in the British Army's Jewish Brigade during the Italian campaign, and in the 1956 war he had commanded an IDF infantry brigade in Sinai. In 1964 he became commander of the Armoured Corps and within the space of a year had achieved a dramatic improvement in the standard of its gunnery. His division consisted of the regular 7th Armoured Brigade (79th Tank Battalion with 66 90mm Pattons, 82nd Tank Battalion with 58 Centurions, an armoured infantry battalion in M3 half-tracks and a reconnaissance squadron); the reserve 60th Armoured Brigade (a tank battalion with 52 Shermans, a light tank battalion with 34 AMX-13s and an armoured infantry battalion); a regular paratroop brigade serving as armoured infantry, supported by a tank battalion partly equipped with upgunned Pattons; and a divisional reconnaissance group which included eighteen upgunned Pattons.

The second armoured division, commanded by Major General Ariel Sharon, was positioned opposite Abu Agheila. In the early 1950s Sharon, a natural swashbuckler, had raised and led a special forces unit and in 1956 he had commanded the paratroop brigade that had secured the Mitla Pass. A strict disciplinarian, he was also a difficult subordinate who was inclined to exceed his orders. Notwithstanding this tendency, he was a good, hard-fighting soldier who could be relied upon to batter his way through any opposition. His division included the 14th Armoured Brigade consisting of one Centurion battalion with 56 tanks and an upgunned Sherman battalion with 66 tanks; two upgunned Sherman companies with 28 tanks that were allocated to the 99th Infantry Brigade, which was to assault the Egyptian trenches at Abu Agheila; and a divisional reconnaissance group reinforced with 20 AMX-13 light tanks.

The third armoured division, commanded by Major General Avraham Yoffe, was positioned midway between Tal and Sharon with the twin responsibilities of stopping lateral reinforcement between the enemy's major defended localities at Gaza and Abu Agheila and preventing intervention by the Egyptian 4th Armoured Division. Yoffe had held a commission in the British Army during World War II and, like Tal, he had commanded an infantry brigade in Sinai during the 1956 war. His division contained the 200th and 520th Armoured Brigades and possessed a total of 200 Centurions.

The war began on 5 June with a series of pre-emptive air strikes by the Israeli Air Force, intentionally timed to coincide with morning rush-hour in Cairo, when Egyptian senior officers would be trapped in dense traffic between their homes and offices. After most of the Egyptian Air Force had been destroyed on the ground, the IAF turned its attention to airfields in Syria, Jordan and Iraq. By late afternoon it had gained complete command of the air and was able to divert squadrons to support the ground fighting.

Codenamed 'Red Sheet', the Israeli offensive in Sinai commenced at 0815, thirty minutes after the first air strikes had gone in. A parachute brigade, supported by AMX-13s, penetrated the Gaza Strip and immediately became involved in heavy fighting with the Palestinians. Simultaneously, Tal's division, spearheaded by Colonel Shmuel Gonen's 7th Armoured Brigade, broke into the flank of the Strip further south and, discounting its casualties, fought its way through Khan Yunis. Gonen's task was to break out along the coast road to El Arish but at Rafah his spearhead was counter-attacked by the Egyptian 7th Infantry Division's armour, led by JS IIIs. Using his Centurions to hold the enemy's attention to their front, Gonen sent off his Pattons in a wide hook through an area of dunes to the east, from which they emerged to fall on the Egyptians' flank and rear. Although the JS III was well armoured and its 122mm gun was capable of defeating both the Centurion and the Patton, it had not been designed for this sort of fast-moving mêlée. The domed turret, like that of the T54/55, gave the loader little headroom in which to work; furthermore, within this cramped space, he was forced to struggle with heavy two-piece ammunition, with the result that the tank's rate of fire was restricted to three or four rounds per minute. Against crews who had been taught speed-loading and who could handle their one-piece rounds in adequate space, such an engagement could have only one ending. By noon the ground was littered with wrecked and burning JS IIIs and anti-tank guns lying shattered in their pits.

Gonen found it significant that both Egyptian divisions had chosen to fight their own battles without any attempt at a coordinated response. Without further delay he ordered his deputy, Lieutenant Colonel 'Pinko' Harel, to collect as many Centurions as possible and press on along the coast road towards the one obstacle remaining in the path of a clean breakthrough, the Jiradi Defile. Hammering along, the crews were uneasy about the prospect, for the defile was over eleven miles long and was known to be held in strength. In fact, covering the eastern entrance alone was an entire infantry brigade, well dug in among minefields covered by anti-tank guns, supported by an artillery brigade with 42 guns and a battalion of 36 Shermans fitted with AMX-13 turrets, many of which were also dug in. Further along the defile and at its western exit were more, though less formidable, defensive positions.

Yet, as the saying goes, who dares wins. The Egyptians, listening to the excited voice on Radio Cairo announcing with more optimism than accuracy that their comrades had captured Beersheba, were quite unprepared for the sudden appearance of the Israeli tanks. Turrets swinging alternately right and left, the Centurions roared the length of the defile, a rain of coaxial machine-gun fire and HE shells keeping most of the defenders pinned down and unable to man their weapons. Having sustained few casualties and little damage, the group, consisting of eighteen Centurions, two Pattons and several half-tracks and jeeps, reached the outskirts of El Arish in mid-afternoon. Harel reported the fact to brigade headquarters from whence it was relayed to an astonished Tal, whose command group had just reached Rafa. Tal, recognising that Harel, now seriously short of fuel and ammunition, would remain in considerable danger unless he could be reinforced, ordered Gonen to reinforce him with the Patton battalion.

This was far easier said than done, for the Egyptians were now fully alert and fiercely determined that the defile would remain closed. An attack along the road, accompanied by a short hook through the soft sand to the south, resulted only in tanks being lost on mines or knocked out by anti-tank guns. The battalion commander, Lieutenant Colonel Ehud Elad, was killed while personally leading the attack. Gonen decided to try again, smothering those positions closest to the road with mortar bombs while the Pattons used their speed and firepower to smash a way through. The result is vividly described by Shabtai Teveth in his book *The Tanks of Tammuz*.

'The time was 1800 hours. The forward tanks were driving at a speed of 45 kph and the distance between each tank was growing. Major Haim [had taken over the battalion on Elad's death] was worried that the Egyptians would be able to pick off each vehicle one after the other. About two kilometres down the road, close to the enemy's artillery positions, he turned left off the road with part of the force and began shooting at the enemy artillery, tanks and anti-tank guns, which were now to his rear. He ordered the rest of the force, which had brought up the rear of the column, to carry on towards El Arish. Anti-tank guns were sent flying and dug-in enemy tanks reduced to flames. When he had a moment Major Haim reported to Colonel Shmuel [Gonen] on the situation and of his intention to destroy the defence zone from the rear.

'"Leave everything, get to El Arish!" Colonel Shmuel ordered.

'Again the distance between the tanks lengthened, with each tank driving through its own stretch of hell and the gunners working like men possessed.'

Suddenly, just as the sun was setting, the Pattons broke out of the defile and found themselves among Harel's Centurions. Many had casualties aboard and all bore scars testifying to the ferocity of the encounter; incredibly, one waddled in with a smashed drive sprocket, lurching wildly from side to side.

Gonen had started to follow the Pattons with his command group and supply echelon but ran into such a hail of fire that he was forced to halt; the defile was closed again. Meanwhile, the 60th Armoured Brigade, commanded by Colonel Men, had been moving along a parallel axis inland. Tal ordered Men to attack the Jiradi position from the south, using his AMX-13 and armoured infantry battalions. The going in the soft dunes, however, was extremely difficult and at 1930 Men informed him that both battalions had stalled short of the objective and were out of fuel. Tal's division was now stationary and dispersed across a wide area stretching from Rafah to El Arish and only prompt, decisive action would get it moving again. He withdrew Gonen's armoured infantry battalion and a Centurion company from mopping-up operations at Gaza, directing his Chief of Staff to lead them personally through the traffic jam which had developed along the coast road. There was no time to be wasted in argument; those trucks which were slow in leaving the highway were bundled off it by the tanks.

With artillery support, the Centurions secured a lodgement in the northern defences of the Jiradi. At midnight, the armoured infantry arrived and, leaving their half-tracks, began fighting their way forward through trenches and bunkers, assisted by parachute flares. The Egyptians fought back stubbornly, but in such close-quarter night fighting their heavy weapons were less effective. As soon as a narrow corridor had been cleared along the road, Gonen set off with his fuel and ammunition lorries and by 0200 was clear of the defile. The Centurion and Patton crews at El Arish immediately began replenishing their vehicles, but if they were hoping for rest and sleep afterwards they were very much mistaken.

Meanwhile, to the south, Sharon's division was heavily engaged at Abu Agheila. Here the Egyptian positions, lying on three successive ridges, had been prepared for all-round defence in great strength and depth. Ostensibly, the position could only be taken by direct assault, and then at heavy cost; that, however, was not the way the Israelis intended solving the problem. During the day Sharon's Centurion battalion, command-ed by Lieutenant Colonel Natke Nir, overran a battalion-sized outpost in a series of sharp actions then went on to cut the roads leading north to El Arish and west to Djebel Libni, shooting up fuel and ammunition dumps on the way; simultaneously, the divisional reconnaissance group and its AMX-13s cut the southern track to Kusseima. The Abu Agheila garrison was now in a situation where it could neither be reinforced nor with-draw. In the afternoon the 14th Armoured Brigade's Sherman battalion closed up to the eastern perimeter, against which the assault was to be delivered that night by three infantry battalions, and began sniping at the enemy's strongpoints.

As darkness fell the 99th Infantry Brigade left the civilian buses in which it had travelled to the front and marched up to its assault start lines, where it deployed with the two Sherman companies it had been allocated. At 2230 Sharon's artillery opened a heavy preparatory bombardment and as the enemy guns began to reply the Israelis played their trump card. Exactly on time, flight after flight of helicopters clattered in to airland a para-chute battalion just behind the Egyptian battery positions, which were quickly stormed. Concurrently, a Centurion-led battlegroup broke through the western perimeter and joined the paratroopers. As soon as the enemy artillery ceased firing, 99th Brigade and its supporting Shermans began fighting their way into the western defences, the tanks pro-viding illumination for their gunners and the infantry with Xenon infra-red searchlights, while the infantry indicated their own progress with coloured light flashers. After several hours of fierce close-quarter fighting the two groups met in the centre of the position just as dawn was breaking. Those Egyptians that could mounted their vehicles and attempt-ed to escape to the south-west, with Israeli tanks in hot pursuit. Abu Agheila, the key to central Sinai, had fallen.

Meanwhile, between Tal and Sharon, Yoffe's division had been grinding its way westwards in low gear. It had traversed an area of dunes that the Egyptians considered to be tank-proof but which the Centurion's cross-country performance and thorough route reconnaissance along the Wadi Haridin proved to be nothing of the sort, although it took nine hours to cover 35 miles. By 1845, having brushed aside such minor opposition as they had encountered, the leading elements of Yoffe's 200th Armoured Brigade were in position covering the track junction at Bir Lahfan, which the Egyptians would have to pass through if they wanted to counter-attack at El Arish. It had, in fact, taken several hours for Mortagy to catch up with the situation, but early that evening he had ordered an armoured brigade and a mechanised brigade to move along the central axis and recapture El Arish in a dawn attack from the south. The Israelis watched them approach the junction in blaze of headlights then, at approximately 2300, opened fire. The headlights were hastily extinguished as the column scattered across the sand, but by now the flames from burning tanks and trucks provided an alternative source of illumination. Although the T55s possessed infra-red night fighting equipment, they were unaware that they were opposed by only twenty Centurions and did not develop an attack, seeming apparently content to engage in a protracted, long range fire fight. This suited the Israelis very well, as it enabled them to maintain their blocking role and simultaneously benefit from their superior gunnery techniques; it was ironic that the only Centurion to sustain serious damage was one which used its Xenon searchlight, and thereafter the use of light projectors was forbidden.

Furthermore, unknown to the Egyptians, Tal had despatched his replenished 7th Armoured Brigade south from El Arish and, having smashed through a defensive position near the town's airfield, this began bringing additional pressure to bear on the Egyptian flank at about 0630. More of Yoffe's division had come up and, as the light strengthened, the IAF came in to strafe and bomb. By 1000 the Egyptians had commenced a rapid and disorderly retreat towards Djebel Libni.

The situation within Mortagy's army on the morning of 6 June can be summarised as follows. At Gaza the 20th (Palestinian) Division was fighting to the bitter end, which could not be long delayed. The 7th Division had been destroyed on the Rafah-El Arish axis, as had the 2nd Division at Abu Agheila. The 3rd Division's camps near Djebel Libni were being savaged by Sharon's and Yoffe's tanks and that formation was pulling back as best it could. On the southern sector, the 6th Mechanised Division and Task Force Shazli had allowed themselves to be intimidated by a single Israeli armoured brigade, Colonel Albert Mandler's 8th with only 50 Shermans at its disposal, with the result that the offensive thrust into the Negev had been abandoned and, at the most critical moment of the battle, several hundred Egyptian tanks were retained to meet a threat that did not exist. In the centre, however, the 4th Armoured Division still lay in the path of the advancing Israelis.

For his part, Gavish had every reason to feel satisfied with the events of the previous 24 hours. However, although the last of the Jiradi's stubborn defenders had been overcome during the night and the 60th Armoured Brigade, having been replenished, had moved forward to El Arish, the very intensity of the first day's fighting ensured that it could not be maintained on 6 June. Having been informed by prisoners that Mortagy had ordered all his troops to pull back and establish a line of defence covering the three passes that led from Sinai to the Suez Canal, the Tassa and Gidi in the north and the Mitla in the south, Gavish flew forward to brief his divisional commanders on the next phase of the battle. His orders were simple: Tal and Yoffe were to smash their way through the retreating Egyptians and seize the passes; Sharon was to complete mopping up around Abu Agheila, then drive the enemy towards Tal and Yoffe so that they would be caught between the hammer and the anvil. As far as possible, most of 6 June was spent in rest and replenishment, although on the coast road a battlegroup commanded by Colonel Israel Granit pushed on for a further 40 miles without encountering serious opposition.

On 7 June the Israelis resumed their breakneck advance. Mortagy, allocated whatever armoured reinforcements could be scraped together in Egypt proper, committed them to the northern sector and did what he could to prevent the complete collapse of his army. The pace of the battle, however, was too fast for him and control slipped from his grasp as one formation after another went off the air.

Between Romani and El Kantara, Granit's battlegroup, which had been reinforced by paratroopers who had driven west at top speed after the fall of Gaza, was temporarily halted by recently arrived enemy armour. While the tanks engaged in a gunnery duel, the paratroopers swung off the road in their jeeps and half-tracks, executing a wide hook onto the Egyptian flank to open fire with their recoilless rifles. Having thus eliminated the last obstacle in his path, Granit pressed on to El Kantara, becoming the first Israeli commander to reach the Canal.

Meanwhile, Tal's division was heading straight for Bir Gifgafa, deliberately seeking battle with Ghoul's 4th Armoured Division. In a stand-up fight lasting two hours Gonen's Centurion and Patton crews demonstrated their superior gunnery, after which the Egyptians pulled off to the south in the fading light. While the tank battle raged, Tal pushed his 60th Armoured Brigade westwards, hoping to fall on his opponents' flank, but the latter broke contact before the move could take effect. At about 0300 the leaguer of the brigade's AMX-13 battalion, located in a hollow beside the road to the Tassa Pass, was entered by two Egyptian trucks with infantry aboard, heading west. Both burst into flames when they were shot up, but while the prisoners were being herded together tanks were heard approaching from the direction of the pass. They were T55s and formed the vanguard of a reinforcement brigade that had just entered Sinai. The guard tanks opened fire at once, their crews watching in horror as the rounds flew off the Egyptians' heavy frontal armour. Since the leaguer was already illuminated by the burning trucks, the Egyptian gunners found plenty of targets for their return fire and soon several AMX-13s, a mortar half-track and ammunition vehicles were also blazing fiercely. Colonel Men, advised of the situation, despatched a Sherman company to the rescue and Tal ordered Gonen to further reinforce this with one of his Centurion companies. Before either could intervene, however, the remaining AMX-13s had scattered into the darkness and begun directing their shots at the enemy's thinner side armour. After the first four tanks in their column had been knocked out in this way, the Egyptians halted and, following some long range sniping, withdrew to the west. Their commander, evidently under orders to buy as much time as possible, deployed his armour in a series of tank ambushes over four miles in length at a point where the metalled road wound through a wide belt of dune country.

Next morning, 8 June, Tal's division resumed its advance on the pass, spearheaded by 7th Armoured Brigade. The leading Centurion company, advancing beyond the bounds permitted by its orders, ran into the first ambush site and sustained some loss before it could be pulled back. Once the situation ahead became clear, Tal resorted to what he described as steamroller tactics. While the Centurions engaged each ambush site in a long range gunnery duel, followed by a simulated frontal attack that absorbed the defenders' attention, the rest of the brigade executed a wide hook through the dunes and fell on the Egyptians' rear. It took the Israelis several hours to fight their way through, by which time the sun was setting and the tanks required replenishment. Political considerations now began to intrude on the battle. Civilian radio programmes picked up on transistor radios indicated that in New York the Egyptian ambassador to the United Nations had requested a ceasefire, and before this could be imposed it was essential that the IDF should be firmly established on the Canal. The IAF had already confirmed that several more Egyptian positions lay ahead, but Tal, beefing up his reconnaissance group with six Pattons and a self-propelled artillery battery, sent it off into the darkness. The reconnaissance jeeps and the tanks worked as a team, the former flashing back signals with their torches whenever they reached a potentially dangerous area, which was then illuminated by the tanks' infra-red light projectors and engaged with direct gunfire if the enemy was present. Most of the Egyptians, however, were already pulling back and little opposition was encountered in the pass. At 0030 on 9 June the reconnaissance group reached the Canal, destroying a guard tank beside which a lonely traffic control sentry mistakenly attempted to direct them across a bridge; replenished, the rest of the division arrived and quickly established contact with Granit's battlegroup to the north.

If anything, Yoffe's exploitation on 7 June was even more dramatic. His axis of advance took him in a south-westerly direction from Djebel Libni through Bir Hasana and Bir Tamada towards the Mitla Pass. It was also the direction in which most of the Egyptians in northern and central Sinai were attempting to retreat, harried constantly by the IAF. This proved to be something of a mixed blessing as, while it further demoralised the enemy and accelerated his disintegration, it also left the roads blocked with a tangle of burning wreckage through which the Centurions had to force their way. Rearguards were little in evidence so that from time to time the tanks, guns blazing, ploughed into the rear of a column. When this happened the Egyptians abandoned their vehicles and fled across the sand, leaving the road still further congested. Progress was so slow that at length Yoffe, realising that he would not reach the pass ahead of the enemy unless drastic steps were taken, despatched a task force, based on Colonel Iska Shadmi's Centurion battalion, to drive through the Egyptians and, stopping for nothing, establish a roadblock at the eastern end of the Mitla. By the time he reached the pass fuel shortage and breakdowns had reduced Shadmi's battlegroup to nine Centurions, two of which were on tow, two infantry platoons and three 120mm mortar half-tracks. Even as the roadblock was being set up, three more Centurions and two of the half-tracks ran out of fuel and had to be towed into position. Disorganised elements of the Egyptian 3rd Infantry, 4th Armoured and 6th Mechanised Divisions and of Task Force Shazli were now converging on the pass and, desperate to escape from the trap in which they now found themselves, they mounted repeated attacks on the tiny Israeli force barring their path. A handful of vehicles broke through before Shadmi's tanks blocked the route with the burning wrecks of those that tried to follow. After this, the growing clutter of their own knocked-out tanks made each attack progressively more difficult for the Egyptians. Shadmi also had a powerful ally in the IAF, which strafed and bombed at will along the three-mile traffic jam that had developed in the approaches to the pass, creating unparalleled scenes of mechanical carnage. During the night most of the Egyptians abandoned their equipment and filtered through the hills on either side of the pass. Yet, it had been a very close-run thing, for when the rest of Yoffe's division broke through at first light on 8 June Shadmi's four remaining Centurions were down to their last few rounds.

After cleaning up the Abu Agheila area, Sharon's division had headed south on the 7th, being joined by Mandler's 8th Armoured Brigade, which had crossed the frontier near El Kuntilla. It had been anticipated that the Egyptian 125th Armoured Brigade, belonging to the 6th Mechanised Division, would put up a fight, yet all its tanks were found abandoned and in working order; the same division's mechanised brigade was encountered near Nakhl and routed with the loss of 60 tanks, 100 guns and 300 vehicles. It remained only for Sharon's troops to drive the remnants of Mortagy's army westwards towards Yoffe and Tal, although for all practical purposes the fighting was over for a while.

The 1967 campaign in Sinai, lasting only four days, cost Israel 275 killed and 800 wounded. Total Egyptian casualties, including 5500 prisoners, were estimated at 15,000; approximately 80% of the Egyptian equipment in Sinai was destroyed or captured, including 800 tanks, 450 artillery weapons and 10,000 assorted vehicles. Not least among the IDF's concepts of armoured warfare to be justified was that of leading from the front at all levels, albeit that several promising middle rank commanders had paid the ultimate price in so doing. Against this, risks were taken that could not have been justified in any other army and although it can be argued that such decisions were made with a full knowledge of the enemy's weaknesses, it was folly to assume that these would never be eradicated. That particular chicken would come home to roost, with consequences that were to shake the entire Israeli nation, during Yom Kippur War of 1973.

Tal was to remain in command of the Armoured Corps until 1969. He served as the IDF's Deputy Chief of Staff in the Yom Kippur War and was then closely involved in the development of Israel's own main battle tank, the Merkava. He is widely regarded as being among the most successful practitioners of armoured warfare since World War II. Sharon again commanded an armoured division during the Yom Kippur War and played an important part in regaining the initiative for the IDF on the Suez Canal front. He subsequently entered politics and became Defence Minister, being noted for his uncompromising views. Yoffe also entered Parliament for a while, then became head of the Israeli Nature Preservation Authority, establishing wildlife reserves throughout the country. It was Gonen's misfortune that he was serving as GOC Southern Command when the Egyptians effected their successful crossing of the Canal in October 1973, and it was inevitable that much of the blame for this and other Israeli reverses should be laid at his door. He was replaced by General Chaim Bar-Lev, but loyally agreed to remain at the front as the latter's deputy.

Most Israelis had a sincere respect for the average Egyptian soldier, but not a lot for his officers, many of whom had abandoned their responsibilities and their men as they sought safety in flight. Well aware of these shortcomings, the Egyptian Army court-martialled no less than 800 senior officers who were either executed, imprisoned for life or dismissed in disgrace. Russian advisers also purged the officer corps of its more privileged members and insisted that the remainder should live and work among their troops. By 1973 the relationship between officers and men, and the general standard of leadership, had improved beyond recognition.

One officer who had escaped the odium of the 1967 defeat in Sinai was General Shazli, who had brought most of his troops home. In 1973, as the Army's Chief of Staff, he was responsible for the detailed operational planning which resulted in the successful crossing of the Canal on 6 October. Believing, correctly, that he could not get sufficient tanks and anti-tank guns across in time to meet the inevitable Israeli armoured counter-attack, he equipped each of his infantry divisions with 314 RPG-7s and 48 portable AT-3 Sagger anti-tank guided weapons. The remarkable success of the latter caused a number of wild-eyed commentators to predict the demise of the tank (similar flawed prophesies had also marked the introduction of the anti-tank gun and nuclear weapons), yet the Israelis quickly discovered that by restoring the proportions of mechanised infantry and artillery in their tank-heavy armoured divisions that the enemy's dismounted Sagger teams could be unsettled or neutralised by sustained fire.

Intervention by the United States and the Soviet Union prevented the Yom Kippur War being fought to a military conclusion. At a heavy cost in lives and equipment, both sides had won and lost territory, but the Egyptians were also satisfied that they had restored their honour. Taken together, this formed the basis for an honourable and lasting peace under the terms of which Israel withdrew from Sinai in exchange for Egyptian recognition and a secure southern frontier.

The Long War
Vietnam, 1946–1975

With the notable exception of the Korean War, the major communist powers preferred to pursue their post-World War II policies of expansion in the Far East by surrogate means, forming well organised 'liberation fronts' from those with nationalist aspirations in colonial territories, training them in the Maoist theory of revolutionary warfare and providing a continuous flow of arms. The areas selected generally consisted of difficult terrain which prevented government troops from employing their armoured vehicles and firepower to best advantage. Being long term, the strategy was not based on the immediate acquisition of territory, but rather on a steady erosion of the enemy's will to fight. This would be achieved by means of raids and ambushes designed to inflict heavy casualties, accompanied by a campaign of murder and intimidation which cowed any opposition among the rural population. In due course, the government troops would find themselves confined to the cities and such fortified posts as they could maintain between them, and in this way large areas of the countryside would pass into the insurgents' control. The next phase would involve penetration of the cities which, through terror and outrage, would be rendered ungovernable. Finally the government would be swept away by a 'popular' uprising. Tactically, a raid on a government post would be used to provoke the despatch of a relief column, the destruction of which was the real object of the operation; afterwards, those involved would melt into the landscape, either escaping through the jungle or, if they belonged to a local guerrilla unit, returning to their fields, where they were indistinguishable from the rest of the population.

Geographically, the French colonies in Indo-China, especially Vietnam, provided an ideal environment for this kind of warfare. Here there existed wide variations in terrain, including jungle, plantations, rice paddy, mangrove swamps, mountains, forested hills and river deltas, all of which presented conventional troops with operational problems. Furthermore, most of the population lived in the countryside and was therefore vulnerable to communist intimidation. Active opposition to the French had actually commenced in 1946 and three years later it began to gather momentum following the communist victory in the Chinese Civil War. The French Army was hamstrung throughout by decisions made in Paris, where political expediency always took priority over military considerations. For example, as conscripts could not be sent to Vietnam, the purely French element of the Army was restricted to the number of units which could be formed from regular volunteers, so that of the 150,000 soldiers serving in Indo-China only 50,000 were French, the balance being made up by 20,000 Foreign Légionnaires, 30,000 Africans and 50,000 Vietnamese. Of these more than half were tied down in static garrisons or employed on convoy escort duties across a vast area measuring approximately 230,000 square miles. The cutting edge of the army's offensive response was provided by eight parachute battalions, for whom there were never sufficient transport aircraft, and seven mobile groups equipped with tanks, armoured cars, half-tracked APCs and LVTs, mainly of World War II vintage, depending on where they were operating.

The French had no recent experience of jungle warfare and, like the British and Americans in 1941–2, they were mentally road-bound. They were also kept permanently short of resources. The only helicopters in Vietnam, for example, were casualty evacuation and supply craft; in 1952 there were just ten of them, although by 1954 the number had risen to 42. For armoured units, already fragmented across the country, the story was much the same. Spare parts took between six and twelve months to reach units, and there was a desperate shortage of armoured recovery vehicles, the work of which had to be undertaken by gun tanks, thus diluting a unit's assets still further.

Despite these disadvantages, numerous defeats were inflicted on the communists, albeit without bringing a resolution of the conflict in sight. In November 1953, the French government announced that it would accept an honourable settlement and delegates from both sides met in Geneva. On 7 May 1954 the position of the communist Viet Minh negotiators was strengthened by the well publicised fall of the French base at Dien Bien Phu. Although the defeat involved just five per cent of the total French strength in Vietnam, the troops lost included the cream of the strategic reserve and this reduced the Army's capacity to mount offensive operations for some time to come. The leader of the Vietnamese communists, Ho Chi Minh, and his field commander, General Vo Nguyen Giap, were naturally inclined to exploit the situation to the full, but they were virtually pushed aside by the Soviet Union and China, who had suffered economic dislocation as a result of the Korean War and the need to support the fighting in Indo-China, and were also worried by the possibility of American intervention. The agreement eventually concluded between Pierre Mendès-France, the French Prime Minister, Chou En-lai, Prime Minister of China, and Vyacheslav Molotov, the Soviet Foreign Minister, provided for the full independence of Laos, Cambodia and Vietnam; the temporary partition of Vietnam along the 17th Parallel pending nationwide elections to be held in 1956; and the withdrawal of French and Viet Minh forces from, respectively, North and South Vietnam.

Ho and Giap had no intention of accepting such a settlement. Having secured their position in North Vietnam, they renewed the revolutionary war process in the south. By October 1957 guerrilla groups, known as the Viet Cong, had begun operating in the Mekong Delta and, with the active support of the North Vietnamese Army (NVA), had established themselves firmly throughout South Vietnam within three years.

While it had inherited some traditions and equipment from the French, the South Vietnamese Army (ARVN) had opted for American military assistance and advice. Despite this, it remained essentially an amateur army attempting to deal with a tough, experienced and very professional guerrilla force, and the situation continued to deteriorate throughout 1961 and 1962. In the latter year, however, it received its first consignment of M113 APCs and wherever these were employed the tactical picture underwent a sudden and dramatic change.

The M113, manned by a crew of two and a ten-man infantry section, was one of the success stories of the long war in Vietnam. Designed by the US Army Ordnance Tank-Automotive Command and manufactured in large numbers by the Food Machinery Corporation of San José, California, the vehicle was constructed with aluminium armour and powered by a 209 hp Chrysler V-8 engine. It was fully amphibious, being propelled by its tracks when afloat, and could cope with almost any sort of going save for soft mud, steep banking and close jungle. Its standard armament consisted of one .50 calibre heavy machine-gun pintle-mounted beside the commander's cupola.

In June 1962 the ARVN committed its first two M113 companies to action in the area of the Mekong Delta. By the end of October they had killed 517 Viet Cong and captured 203 at a cost of only four dead and 13 wounded, as well as penetrating numerous areas in which the communists had previously felt secure. After this encouraging start six more such units were raised and designated Mechanised Rifle Squadrons, being assigned to the Army's four Armoured Cavalry Regiments, the establishment of which also included one tank squadron with M24 Chaffee (later M41 Walker Bulldog) light tanks and a reconnaissance squadron with M114 command and reconnaissance vehicles.

By and large, the mixed fortunes of the ARVN have tended to obscure some of its better features. As far as the M113 squadrons were concerned it deserves the credit for a number of imaginative innovations in both the technical and tactical fields. In the Mekong Delta, where canals and branches of the river formed natural obstacles, a variety of ingenious techniques were developed for crossing these and recovering bogged vehicles; these included block-and-tackle, ground anchors, linked towing, small fascines and flat aluminium push-bars which, when used in pairs, could also be employed to create a path over soft going in much the way as sand channels in the desert. American tactical doctrine, like that other conventional armies, favoured APCs dropping their infantry on the assault start line, after which the attack would be made on foot with fire support. The ARVN infantry, however, preferred to fight mounted, partly to maintain the momentum and shock of an armoured attack, and partly to reduce casualties in country where the enemy habitually protected his positions with mines and booby traps; only when a communist position had been overrun did the infantry dismount and clear the area. There were, nevertheless, occasional setbacks such as that at Ap Bac in January 1963 when a high proportion of M113 commanders were killed while manning their pintle machine-guns. This led to the fitting of .30 calibre machine-guns on the sides of the vehicles and the protection of all gun positions with armoured shields. These modifications were later standardised by the US Army and in this form the M113, which now possessed greater firepower than many light tanks at the beginning of World War II, was known as the ACAV (Armored Cavalry Assault Vehicle).

There is no doubt that the sudden appearance of M113s in areas where armour had never operated previously came as a most unpleasant shock to the communists. They reacted by digging tank traps in roads and mining known river and canal crossing points. Various shaped charge anti-tank weapons were hurriedly distributed, but it took almost two years before they became standard issue at company level. Best known and most formidable of these was the Soviet RPG-7 rocket-propelled grenade. In his book *Vietnam Tracks*, Simon Dunstan makes the point that the basic inaccuracy of the weapon resulted in only one hit for every eight or ten rounds fired, and that for a hit to be effective it had to strike as near as possible to the critical angle of 90 degrees. M113s and ACAVs were, on average, only penetrated by one in every seven hits, and each penetration caused an average of one casualty among the occupants; only one vehicle in seven was actually destroyed by penetration. Tanks, having angled armour, were at lesser risk than the flat-sided M113s, and possessed an additional advantage in that their stowage bins and spare track links placed on vulnerable areas provided a form of spaced armour that dispersed the force of the explosion before it reached the body of the vehicle. Mines present-ed a greater danger to the occupants of M113s as they could blow in the belly and cause havoc within. The floor of the crew compartment was therefore covered with a layer of sandbags but even so most infantrymen preferred to ride on the roof of the vehicle.

In December 1961 the ARVN had commenced airmobile operations using American helicopters, and the success of these, supplemented by the local victories won by the South Vietnamese armour, confirmed that Giap's concept of revolutionary war-fare, so successful against the ill-equipped French, had failed to keep pace with the latest developments in military technology. Despite this, the situation within South Vietnam continued to deteriorate steadily. At the end of 1961 there were 15,000 American advis-ers in the country, rising to 23,000 three years later. In August 1964 an abortive attack on US warships by North Vietnamese torpedo boats led to immediate counter-strikes and the passing of a Congressional resolution permitting President Lyndon Johnson to take whatever steps he considered necessary to safeguard American lives and prevent fur-ther aggression. In October and November Viet Cong attacks killed four American advis-ers and wounded 52 more. On 8 March 1965 the arrival of two Marine Corps battalions at Da Nang airbase marked the formal entry of the United States into the ground war. By the end of 1968 there were 536,000 American troops deployed in support of the ARVN, plus contingents from Australia, New Zealand, Thailand and South Korea.

The Americans had serious reservations about the use of tanks in Vietnam, based upon an incomplete understanding of the fate of the French Mobile Group 100. In fact, despite its title, this formation consisted of vulnerable lorried infantry and only one squadron of obsolete Stuart light tanks. Even so, it had pursued an active career in enemy-dominated country for six months before it was destroyed in a series of ambushes.

The deployment of American tank units to Vietnam might well have been delayed indefinitely had not the Marines brought their own organic armour with them. The M48A3 Patton medium tank quickly demonstrated that it could be applied very successfully to numerous tactical problems; its power and weight, for example, enabled it to crush its way through all but the closest jungle and its 90mm gun could not only be used at close quarters with devastating effect against the enemy's bunkers, but also fire a canister round known as Beehive, containing over five thousand metal flechettes that could be used against personnel, or to clear dense vegetation from the path of the vehi-cle, or even to detonate mines. Given the type of war they were involved in, the crews stowed very little armour piercing ammunition, preferring instead a majority of canister rounds plus appropriate proportions of high explosive and white phosphorous. The M48A3 also equipped the US Army's medium tank units and was later issued to the ARVN. In February 1968 the Australian contingent was joined by a squadron of Centu-rions, which possessed an even higher immunity to the RPG-7 than the Pattons, notwith-standing the fact that their side bazooka plates were removed to avoid compaction of earth and vegetation in the running gear, and were equally good at breaking trails through the jungle.

In January 1969 a new and unproven American light tank, the M551 Sheridan, armed with a 152mm gun/missile launcher, was committed to the fighting in Vietnam. It was heartily disliked by its crews for a number of reasons. Tactically, it was vulnerable to mines, was unable to break jungle as well as the medium tanks, and its bagged ammunition charges caught fire and exploded at once if the vehicle was penetrated. Technical-ly, it suffered from too many mechanical, electrical and gunnery faults to be considered in any sense reliable, although many of these were cured in time. On the other hand, the 152mm Beehive round quickly acquired a fearsome reputation; in one engagement no less than 36 of the enemy were killed in seconds by two such rounds.

For riverine operations and amphibious landings along the coast the Marine Corps used its LVTP-5s amtracs, and in this role they were perfectly satisfactory. How-ever, attempts to use them as APCs during operations inland were less successful as their fuel tanks, containing 456 gallons of high octane petrol, were located beneath the floor and would explode in a gigantic fireball, killing everyone within, if a mine was struck. Because of the risk, Marine riflemen preferred to ride on the vehicle's roof and often con-structed sandbag parapets around the edge.

By June 1969 the Americans and their allies in Vietnam could deploy over 600 tanks and 2000 armoured vehicles of various types, including the ubiquitous ACAVs and numerous self-propelled artillery weapons. This was not greatly superior to the number of AFVs used by the French, although the latter had been compelled to disperse theirs across an operational area three times the size. Two basic unit organisations were employed by the US Army, although in practice there were wide variations. The first was the tank battalion, containing a headquarters company, a service company and three tank companies, each with a company headquarters and three five-tank platoons. The second was the divisional armoured cavalry squadron, consisting of a headquarters troop, an air-mobile troop with helicopters, and three armoured cavalry troops each with a troop head-quarters and three armoured cavalry platoons with tank and ACAV sections. There was also one larger formation, the 11th Armored Cavalry Regiment, consisting of a head-quarters troop, an airmobile troop and three armoured cavalry squadrons, each contain-ing three armoured cavalry platoons with ACAVs, a tank company and a howitzer battery. The terms squadron, troop and platoon used by the US Armored Cavalry equat-ed with regiment, squadron and troop in the Australian Army and the ARVN.

The principal roles in which armoured troops were employed included search-and-destroy missions, rapid reaction, line of communications security and perime-ter defence. The search-and-destroy concept was very similar to the British cordon-and-search operation, although for various reasons the term became politically unacceptable and was replaced by reconnaissance in force. After the suspect area had been surrounded and all probable lines of escape sealed by cut-off parties or artillery, it was subjected to one or more sweeps that drove the enemy towards the stop line and revealed his positions. Whereas formerly it had been the infantry who led sweeps in close coun-try, in Vietnam it was the tanks which took the lead wherever the going permitted. The reasons for this stemmed from the enemy's use of anti-personnel mines and booby traps which included pits containing spikes that would impale a boot and trip wires that would detonate an explosive device or bring down a large, heavy, swinging mud ball pierced with blades that would mangle anyone in its path. As the tanks broke new trails through the forest, exploding anti-personnel mines and triggering booby traps, they would blast the scrub with canister and machine-gun fire, simultaneously clearing the vegetation and eliminating any opposition in their path. The infantry, still aboard their APCs and ACAVs, would follow behind. When the enemy's bunkers were reached the tanks would engage them with high explosive then crush them beneath their tracks. If the enemy chose to make a determined stand the tanks and ACAVs would continue to engage him closely while additional artillery and air strikes were laid on and fresh troops were lifted into the area by helicopter, a technique known as Pile On. The final phase of the operation would consist of a dismounted sweep by the infantry, who would search the position for prisoners and intelligence material and destroy captured weapons and equipment.

Rapid reaction involved a force of tanks and/or APCs remaining on standby within a fortified base, ready either to go to the assistance of a road convoy or patrol that had run into trouble, or indeed to react to any enemy contact in the area.

Line of communication security, accounting for approximately one-third of all armoured operations in Vietnam, consisted primarily of route opening and convoy escort. These were extremely important duties as one of the principal aims of the communists was to deny their opponents use of the roads between garrisons and supply bases; control of the roads also enabled the communists to levy contributions from civilian traffic, and in so doing emphasise the extent of their grip on the surrounding countryside. Ambush sites were approximately half a mile in length, the ambush usually being initiated by the explosion of remotely controlled mines which halted the first and last vehicles in the column. Heavy fire was then opened with automatic weapons, mortars, RPG-7s and recoilless rifles. Those of the vehicle occupants who attempted to take cover at the roadsides or in ditches found them littered with anti-personnel mines, booby traps and short, razor sharp bamboo stakes known as panjis. The counter-ambush technique involved tanks, ACAVs and APCs swinging alternately right and left to the roadside and opening fire with every available weapon for what was known as the 'mad minute.' The effect of this was to strip the undergrowth in which the guerrillas were concealed, enabling gunners to identify their targets. Whenever possible, the leading AFVs would drive through the ambush and 'clover leaf' round onto the enemy's rear. If necessary, the convoy escort commander could call for assistance and the Pile On technique might be activated. In the worst infested areas artillery and air strikes might be used to clear roadsides ahead of a convoy. Most ambushes were set at night and, to discourage the enemy, groups of AFVs would make irregular but continuous high speed patrols, known as 'thunder runs,' along the more important roads, shooting up the verges and likely ambush sites.

When armoured units were not actively engaged in operations, and especially during the wet season, their vehicles were incorporated into the perimeter defence system of whichever base they were operating from, with interlocked arcs of fire. The vehicles and their crews were dug in to protect them from the rocket and mortar attacks which usually preceded a communist attack and chain link fencing was used as a defence against RPG-7s. After dark, the tanks' Xenon infra-red light projectors provided a vital source of illumination beyond the perimeter when it was required.

These roles dovetailed into a continuous all-arms prosecution of a war in which the Americans accepted Giap's own concept that the terms front, flank and rear meant nothing and that the only object was to bring the enemy to battle and to inflict crippling casualties upon him. To this they brought a mobility and firepower which had been undreamed of in the days of the French and soon the Viet Cong were sustaining such heavy losses that Giap was forced to commit more and more of his regular NVA units to maintain the struggle. Two examples of battles in which the use of armour had a critical effect will suffice.

In the summer of 1966 Major General William E. DePuy's US 1st Infantry Division was cooperating with ARVN units in Long Binh province, to the north of Saigon and on the border between South Vietnam and Cambodia. This area formed part of a long established communist stamping ground known as War Zone C, and even as the 1st Division was fighting to reopen Route 13, the main highway between Saigon and Loc Ninh, the 9th Viet Cong Division was concentrating for an assault on the provincial capital of An Loc. On 8 June part of the divisional armoured cavalry squadron, 1st Squadron 4th Cavalry, commanded by Lieutenant Colonel Leonard L. Lewane, was heavily ambushed near Ap Tau but succeeded in holding its own during a stiff fight lasting four hours. Subsequent analysis of the action resulted in improved methods of coordinating artillery and air support, and plans for rapid reinforcement by airmobile infantry who would counter-attack the ambush force from the flank as well as block the enemy's escape route. These methods were put into effect when two of 1st Squadron's troops were again ambushed near Srok Dong on 30 June, with the result that the survivors of the Viet Cong unit responsible fled across the border into Cambodia leaving 270 of their comrades dead around the ambush site, together with most of their heavy weapons.

Nevertheless, DePuy was convinced that the communists would, for the sake of face, continue to mount further ambushes and he believed that these could be turned to his advantage. The most probable site was examined and infantry and artillery units were positioned on suitable landing zones within a few miles, ready to intervene as soon as the ambush was sprung. To encourage the enemy, deliberately careless talk let it be known that a convoy consisting of a bulldozer, several supply trucks and a small armoured escort, would be travelling from An Loc to Minh Thanh on the morning of 9 July, a piece of intelligence which the civil population's Viet Cong sympathisers could be relied upon to relay to their masters.

At 0700 on 9 July the convoy, codenamed 'Task Force Dragoon', left the perimeter at An Loc and travelled south on Route 13 for one and a half miles before turning right onto the secondary unsurfaced road that ran in a south-westerly direction towards Minh Thanh. Far from being the soft target suggested by rumour, it consisted of the 1st Squadron's B and C Troops reinforced by B Company 1st/2nd Infantry in APCs. The column moved slowly and with the customary caution and by 1025 it had reached a bridge known as Checkpoint Dick, some 1500 yards short of the ambush site. Here it halted and went through the standard drill of checking the bridge for explosives while artillery and air strikes hit the trees and scrub beyond. The Viet Cong were also familiar with the drill and made no response. With 1st Platoon C Troop leading, the column crossed the bridge and continued along the road, now painfully aware that it was driving deeper and deeper into the ambush site. Ahead, a helicopter gunship was strafing the next checkpoint, codenamed Tom, but once again the Viet Cong remained silent.

At about 1100, when the head of the column was approximately halfway between Dick and Tom, two groups of Viet Cong were observed running across the road. The leading tank promptly let fly with canister and almost immediately an extremely heavy fire was directed at C Troop from the north side of the road. In quick succession the leading tank commander was killed, the commander of 1st Platoon was wounded and two of his APCs were set ablaze. Now commanded by their sergeant, the rest of the platoon, consisting of two tanks and four M113s, fought back with all their weapons, including an M132 flamethrower. C Troop's 2nd Platoon quickly closed up and herringboned to add the weight of its return fire, but the troop's 3rd Platoon was so heavily engaged that it was unable to close a 300 yard gap in the column. The task force commander immediately ordered B Troop forward into the ambush site so that its vehicles became integrated with those of C Troop.

A contact report had been transmitted as soon as the first shots had been fired. Soon the enemy was also being hammered by artillery and air strikes, but the furious

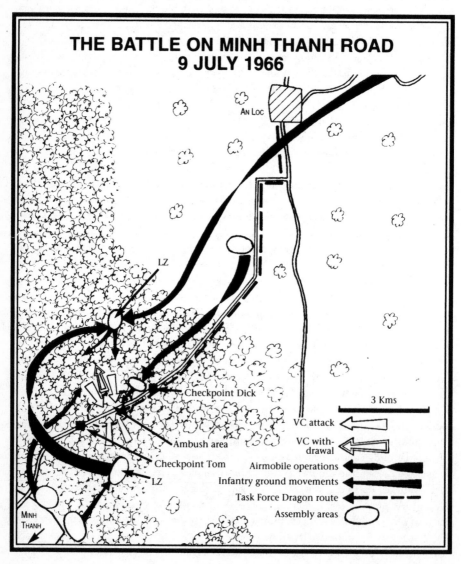

THE BATTLE ON MINH THANH ROAD
9 JULY 1966

An Loc

LZ

Checkpoint Dick

3 Kms

Ambush area

Checkpoint Tom

LZ

Minh Thanh

VC attack

VC with-
drawal

Airmobile operations

Infantry ground movements

Task Force Dragon route

Assembly areas

exchange of fire continued unabated for the next 45 minutes. At the end of this peri-od C Troop's Pattons had expended more than half their supply of canister and its APCs were running out of .50 calibre ammunition. By now, however, the reaction force of airmobile infantry units had been lifted from its staging areas and was bringing pres-sure to bear on the flanks of the enemy ambush force, as well as moving into blocking positions behind the Viet Cong. The scale of artillery and air support had also been stepped up and, as B Troop seemed capable of retaining control of the situation, the task force commander ordered C Troop back to Checkpoint Dick to replenish its ammunition.

After a further 30 minutes of heavy fighting the Viet Cong broke and fled, only to be intercepted by the blocking force and cut down by artillery and helicopter gunships. The following day a sweep of the area by dismounted infantry discovered small groups still attempting to escape. The Viet Cong sustained the loss of 240 killed, plus an unknown but undoubtedly large number of wounded who had been removed before the trap closed; eight prisoners and a quantity of weapons were also captured. As a direct result of this series of engagements the 9th Viet Cong Division was so serious-ly mauled that it not only had to abandon its plans to seize An Loc, but also to relin-quish its grip on the strategic Route 13. Significantly, the Battle on the Minh Thanh Road demonstrated that whereas in conventional warfare infantry would be used to hold the enemy while armour manoeuvred to encircle him, in Vietnam the reverse was true.

In numerous actions of this type, plus the corps-sized search and destroy oper-ations codenamed 'Cedar Falls' and 'Junction City', the Viet Cong began to suffer severe-ly. To justify their horrific losses and save face, local communist commanders informed their superiors that not only were they inflicting even heavier casualties in men and materiel on the Americans and their allies, but also that large areas of South Vietnam were ripe for a popular rising. Although this ran contrary to the instincts of Giap and his Chinese backers, the final phase of revolutionary warfare was activated in what became known as the Tet Offensive of January 1968. The result was a shattering defeat and the virtual destruction of the Viet Cong. From this point onwards the burden of the com-munist war effort was shouldered by the North Vietnamese Army, and it is against this background that the next example is set.

In April 1968 the bulk of the Australian Task Force began moving north from its base at Nui Dat in Phuoc Tuy province to an area on the boundary between Bien Hoa and Binh Duong provinces, approximately twenty miles north-east of Saigon. Intelli-gence reports indicated that the NVA intended moving south through the area with a view to attacking targets closer to the capital. The Australians' orders, therefore, were to establish themselves in fire support bases either side of Route 16 and prevent this.

FSB Coral was occupied on 13 May and that night the NVA mounted an attack on the position, a clear indication that the Australians were interfering with their plans. The attack was contained with difficulty and was repeated in much greater strength two nights later. By now the defences were stronger, but had not the M113 APCs of A Squadron 3rd Cavalry Regiment, commanded by Major J. D. Keldie, been present, it is probable that the base would have been overrun. During the fighting the carriers had pro-vided a vital source of firepower, had brought badly needed ammunition to threatened areas of the perimeter, and evacuated wounded men under fire. When the enemy finally withdrew, one troop pursued him so closely that, uncharacteristically, he abandoned his wounded and numerous weapons.

Nevertheless, the result had been too close for comfort and for additional security the Centurions of C Squadron 1st Armoured Regiment were ordered forward from Nui Dat to FSB Coral. On 24 May a further fire support base, Balmoral, was established three miles north of Coral. The following day a four-tank Centurion troop, commanded by 2nd Lieutenant J. M. Butler, was sent on to Balmoral and was integrated into the perimeter defences. At 0400 next morning the enemy launched a battalion-sized attack which the tanks assisted in repelling with canister fire.

On their way to Balmoral, Butler's troop and the accompanying infantry company had come across a well developed enemy position on the edge of thick jungle but had skirted this after a sharp fire fight. On the morning of 26 May four more Centurions, commanded by Lieutenant G. M. McCormack, and B Company 1st Royal Australian Regiment under Major A. W. Hammett, set off from Coral to subdue this. The advance into the scrub was made with the infantry leading until the rattle of small arms indicated contact with the enemy. The tanks then edged forward slowly in parallel, being hit repeatedly by noisy but ineffective RPGs, but did not open fire until they were certain that they had passed through all of their own riflemen. They then used canister and machine-gun fire to strip the foliage around a arc of 180 degrees, revealing numerous bunkers. The Australians, recalling the value of armour piercing shot in smashing up bunker timbers during the Papua/New Guinea campaign, stowed a higher proportion of AP than the Americans, and this they proceeded to use to good effect, grinding forward slowly to crush the shattered structures beneath their tracks while the infantry followed closely behind, mopping up with rifle, grenade and flamethrower. Altogether, fourteen bunkers were destroyed and their occupants killed before the Australians withdrew without having sustained a single casualty; all the tanks showed signs of external damage but none was immobilised.

The Australians had clearly become an extremely painful thorn in the communists' side. Evidently, orders were issued that they should be eliminated, whatever the cost, for at 0100 on 28 May an attack in regimental strength was mounted against Balmoral and pressed home with the utmost determination for several hours. With the assistance of artillery firing from Coral and nearby American bases every assault was broken up. During the night each of the four tanks expended fifty or so canister or HE rounds, the grisly results being piled up in the wire and far beyond. Many Australians were sickened by the slaughter and the enemy's dead were quickly buried in bomb and shell craters by a bulldozer.

On 30 May the NVA came close to exacting their revenge. During the morning A Company 1st Royal Australian Regiment, commanded by Major Ian Campbell, left Coral aboard a troop of 3rd Cavalry APCs to patrol an area of dense jungle approximately 3000 yards east of the base. Arriving at the edge of this the riflemen dismounted and continued on foot. They soon became engaged in a heavy fire fight with a bunker complex and were pinned down. Casualties began to mount as the enemy began to work round the company's flanks. The NVA brought the fight to very close quarters, employing the tactic known as 'hanging onto the enemy's belt' or 'hugging,' which prevented

Campbell calling in artillery or air support. Now in potentially serious trouble, he requested tank assistance.

As luck would have it, the tank troops at Coral were having a maintenance day, although as a safety precaution not all the Centurions had been stripped down simultaneously. With a composite troop of three tanks, McCormack roared off to the rescue. The APCs indicated that they would show the way but were neither powerful enough nor heavy enough to crush a path and their bows simply rode up on the compressed vegetation until traction was lost. McCormack, now reduced to his own and Sergeant P. J. Reeves' tanks after a radio failure in the third Centurion, decided to break trail himself while the APCs followed, covering his rear. As Australian casualties were still on the ground he had to proceed with extreme caution but at last reached a point where he received clearance to open fire. Swathes of canister cut through the jungle, revealing bunker entrances and fire slits. While the tanks edged systematically forward, blasting bunkers with high explosive and crushing them with their tracks, the riflemen began loading their own casualties aboard the APCs. At one point McCormack could hear an enemy machine-gun firing nearby and asked the infantry to pinpoint its position. 'You're sitting on it!' said the voice in his headphones. Leaning over the edge of the cupola his saw that the tank's tracks were resting beside a bunker entrance, into which he dropped two grenades; after that, the weapon fell silent. One position, though hit several times, opened fire repeatedly until McCormack slammed an armour piercing round through its roof timbers. These collapsed, revealing a zig-zag communication trench along which the enemy were replacing and evacuating their casualties.

The tanks had now won Campbell sufficient elbow room for him to disengage his platoons and pull them back to the APCs. Simultaneously, they had also cut down enough jungle for helicopter gunships and artillery to intervene, not least against the escape routes on which the enemy was now converging. Having eliminated the heart of the bunker complex, the tanks covered the infantry withdrawal and then pulled back themselves along the paths they had carved. A patrol visiting the area two days later found that the enemy had gone for good, his morale so shattered that he had never returned to recover his dead and their weapons.

On 6 June, its mission completed, the Australian Task Force moved back to Phuoc Tuy province. The lowest possible estimate of enemy killed in the fighting around FSBs Coral and Balmoral was put at 267; Australian dead amounted to approximately 10 per cent of this figure. Most NVA soldiers had carried at least two RPGs, an indication of the importance attached to the elimination of Allied armour, although in the event they had also been used against trenches and anything else their owners fancied. Several prisoners were taken and their interrogation provided a number of interesting points. They were very young, being aged between 16 and 18, and at first they had entered the fight full of idealism, having been told that they were about to take part in the final decisive battles around Saigon. They were bitter about the lies they had been told, that their training had left them unprepared for the sheer volume of firepower they had been required to face, and especially that the lives of their friends had been so ruthlessly squandered.

Herein lay the nub of the problem. Giap was able to continue selling the lives of his soldiers cheaply because the truth was carefully concealed from the North Vietnamese public, who were simply told that their young men were 'fighting in the South.' He was also aware that the American public was becoming intolerant of its own continuing if far lower casualties and that the United States would withdraw if an apparently honourable settlement could be reached. Negotiations, which began in January 1969, were accompanied by a steady reduction in the numbers of American and Allied ground troops serving in Vietnam by 12 August 1972 none were left. By the end of August 1973 the American process of disengagement had been completed. In March 1975 the South Vietnamese Army, thinly spread, lacking adequate air cover, spares, ammunition, fuel and resources of every kind, suddenly collapsed and by 30 April the communist victory was complete. Significantly, Giap's final offensive employed the methods of conventional rather than revolutionary warfare and was spearheaded by armour.

Green Screen Warfare
Operation 'Desert Sabre', Kuwait, February 1991

I n 1979 the Shah of Iran, regarded by the West as a stabilising influence in the volatile politics of the sensitive oil-rich Gulf area, was overthrown by a revolution which was rapidly dominated by Islamic Fundamentalists. As a result of this the efficiency of the Iranian Army, which was equipped to a high standard with Western weapons, deteriorated very quickly when its officer corps was ruthlessly purged or fled abroad.

In September 1980 Saddam Hussein, the dictator of Iraq, hoping to profit from his neighbour's weakness, launched an invasion of Iran. His aims are still obscure, although he almost certainly intended annexing the oil-bearing border province of Khuzistan. Having been promised by Iranian exiles that the revolutionary regime would collapse under pressure, he anticipated an easy victory. An ineptly executed Iranian counter-attack with an armoured division was indeed defeated but, to Saddam's surprise, most Iranians loyally rallied behind their government. Unable at first to conduct sophisticated operations, the Iranians resorted to brutal human wave tactics and, contrary to most expectations, these succeeded not only in recovering most of the territory which had been lost but also carrying the war into Iraq. Thereafter, although the constantly expanded Iraqi Army enjoyed an undoubted material superiority, neither side apparently possessed the ability to inflict a decisive defeat on the other. By the time a ceasefire had been concluded in August 1988 the status quo ante bellum had more or less been restored at the cost of half a million lives.

The war left Saddam with the fourth largest army in the world, numbering approximately 955,000 men, equipped with Soviet weapons including some 5500 main battle tanks (MBTs), of which 1000 were modern T72s, 6000 APCs and about 3500 artillery weapons, most of which were towed. Also included in his armoury were a number of Scud surface to surface missiles which rumour suggested were capable of delivering a biological or chemical warfare payload, with a nuclear capability hinted at in the near future; and since Saddam had no qualms about using gas against Kurdish dissidents in the north of his country, the threat posed by these weapons had been one of the factors that had finally brought Iran to the negotiating table.

Having failed to achieve any tangible benefit from his war with Iran, Saddam turned his attention to the tiny but prosperous Gulf oil state of Kuwait, which his troops overran on 2 August 1990. Obviously, so blatant a flouting of international law could not be tolerated, especially as the result left an unacceptably large proportion of the world's oil resources under the personal control of an unstable dictator. While the United Nations passed a series of resolutions demanding an Iraqi withdrawal from Kuwait, King Fahd of Saudi Arabia, believing that his country would become the next victim of Saddam's ambitions, requested the active assistance of the United States. In an operation with the overall codename of 'Desert Shield', the first American troops reached Saudi Arabia on 9 August and were followed by contingents from many other nations, including the United Kingdom. As month succeeded month, both sides continued to build up their strength until 350,000 Iraqi troops equipped with 4300 tanks faced a 665,000-strong

UN Coalition army with over 3600 tanks across the Saudi/Kuwaiti border. When Saddam promised 'the mother of battles' if the Coalition attempted to interfere with his hold on Kuwait, many believed him. There were, however, other factors to consider.

Saddam's psyche was far from simple and defied straightforward analysis, its most obvious trait being the ability to expose and play upon his opponents' fears, including the American government's post-Vietnam reluctance to incur casualties and the West's horror of chemical weapons. For all this, neither he nor the members of the Ba'ath Party who formed his Revolutionary Command Council displayed any grasp of strategy and, rejecting such advice as was deferentially offered by the chiefs of Iraq's armed services, they planned a campaign that was based on a series of flawed premises. The first was that, having received a political mandate for the liberation of Kuwait, the Coalition Army was bound to restrict its operations to that country. The better part of 43 divisions were therefore stuffed into Kuwait, where they proceeded to dig themselves into static positions sited in depth. Saddam anticipated that the Coalition would sustain serious casualties as it fought its way through these and that in this weakened state it would be unable to withstand a counter-attack by his Republican Guard. As a result of this deployment, very few Iraqi formations were left guarding Iraq's frontier with Saudi Arabia to the west of Kuwait. The second was a failure to understand the basic principle that the tank is an offensive weapon and that, used in any other way, its worth is devalued. Given that war had become inevitable, Saddam's failure to use his ample armour aggressively while the Coalition forces were assembling forfeited the best opportunity to inflict casualties and so influence American political opinion; during the war itself the Iraqi Army's only offensive operation was a thrust at the coastal town of Khafji and, although this was ultimately to prove abortive, it succeeded in generating quite disproportionate alarm during its early stages. There was, too, a tendency for the Iraqi hierarchy to become bemused by its own propaganda and underestimate the potential of the Coalition army. This was extremely dangerous, for the American and British contingents had spent the past 40 years preparing to meet an onslaught by the recently collapsed Soviet Union and its satellites, the result being that their accumulated experience and standardised procedures enabled them to operate with a speed that the Iraqi Army was unable to match.

Some Western media commentators, with a naivety reminiscent of the Vietnam War, also swallowed Saddam's propaganda and unwittingly played his game for him. On television, talking heads spoke in awe of the Iraqi Army's size without analysing its efficiency and, referring to the recent war with Iran, described it as battle hardened without examining its performance. In this jargon riddled world of instant opinions, the Republican Guard quickly became elevated to the status of bogeymen. It was referred to as an 'élite,' which might have been justified in the sense that it was politically loyal to Saddam and therefore armed with more modern equipment than the rest of the army; but to describe it as 'crack' was hardly appropriate for a formation whose primary purpose was to repress Saddam's enemies within Iraq itself.

Nevertheless, it was immediately clear to General Norman Schwarzkopf, US Army, the Commander in Chief of the Coalition forces, that whatever the flaws of the Iraqi Army, and however modest its record of success, it still possessed the capacity to inflict unacceptably high casualties, especially as the strength of his own ground troops had not reached the 3:1 ratio considered necessary for offensive operations, let alone the

5:1 ratio needed for an attack on highly developed defensive positions. In the air, how-ever, the Coalition possessed a decisive superiority and Schwarzkopf decided that, prior to initiating the major land battle, he would use this to mount a prolonged preparatory air offensive with object of obtaining complete command of the skies, wrecking the enemy's command, control and logistic infrastructure, and finally destroying the Iraqi Army's will to fight by sustained battering with means that varied from laser-guided pre-cision bombing to explosive carpets laid by B-52 strategic bombers. Under the codename of 'Desert Storm', the offensive began during the night of 16/17 January 1991 and with-in four days achieved its first objective when the Iraqi Air Force, unwilling to engage in air combat yet unable to protect its aircraft on the ground, fled to Iran, where it was interned. This left the ground troops to endure the full fury of the onslaught day after day, week after week.

While the air attacks continued, Schwarzkopf completed his preparations for the ground offensive, the aim of which was nothing less than the complete destruction of the Iraqi forces in Kuwait. To students of desert warfare his plan contained recognis-able elements of Allenby's at Megiddo in 1918, or O'Connor's at Sidi Barrani and Beda Fomm in 1940/41; to those more familiar with remoter periods it bore similarities to that of Hannibal at Cannae in that his dispositions reflected the equipment, tempera-ment and abilities of the many diverse nationalities that made up the Coalition Army. To reinforce Saddam's conviction that the Coalition would fight the land battle in Kuwait, Schwarzkopf kept a 17,000-strong US Marine amphibious force lying offshore in the Gulf, indicating that a landing was probable behind the enemy's main defence line, with the result that several Iraqi divisions were deployed to meet the threat. An attack, delivered by armoured and mechanised formations provided by Syria, Egypt, Saudi Arabia, Kuwait, Oman, Qatar and the United Arab Emirates, together with two US Marine Divisions and a brigade from the US 2nd Armored Division, would indeed be mounted along the southern frontier of Kuwait. While this was expected to achieve a breakthrough followed by an exploitation in the direction of Kuwait City, its primary purpose was to divert Iraqi attention from operations taking place further to the west. Here, beyond the point where the Kuwaiti border swung north, the US VII Corps, con-sisting of the British 1st Armoured Division, the US 1st and 3rd Armored Divisions, 1st Armored Cavalry Division, 1st Infantry Division (Mechanised) and the 2nd Armored Cavalry Regiment, was to advance north into Iraq then swing east across the Wadi al-Batin into Kuwait and engage the enemy's armoured reserve before it could intervene in the fighting in the south. Further west still the US XVIII Airborne Corps, consisting of the French 6th Light Armoured Division, the US 82nd Airborne Division, 101st Air-borne Division (Air Assault), 24th Infantry Division (Mechanised) and the 3rd Armored Cavalry Regiment, was to employ its ground and air mobility to the maxi-mum, advancing quickly through lightly defended Iraqi territory to establish a blocking position in the Euphrates valley, thus trapping Saddam's army in Kuwait inside a strate-gic pocket.

With the exception of the Syrians, who retained their Soviet equipment, as did some Egyptian units, the Arab contingents were armed with Western weapons, notably the American M60 Patton and French AMX-30 MBTs, both of which mounted a 105mm gun; the latter, lightly armoured but fast, also equipped the French 6th Light

Armoured Division. The cutting edge of the American and British armoured divisions were, respectively, the M1 Abrams and the Challenger MBTs, the former mounting a 105mm gun and the latter a 120mm gun; both designs were protected by Chobham armour which incorporated a ceramic layer capable of absorbing the jet of intense heat generated by shaped charge weapons, supplemented in certain areas by reactive armour which dispersed their effect with a counter-explosion.

Both sides, of course, possessed night fighting equipment as well as NBC over-pressure defence systems, but in other respects the American, British and French tank crews, having spent so long preparing for a major war in western Europe, were technically a full generation ahead of the Iraqis. In addition to a highly sophisticated command and control apparatus, they possessed navigational aids, thermal image gunsights, laser rangefinders and fire control computers. Such widespread use of electronic equipment, much of it beyond the imagination of their World War II counterparts, actually presented little difficulty to a generation which had grown up with home computers and

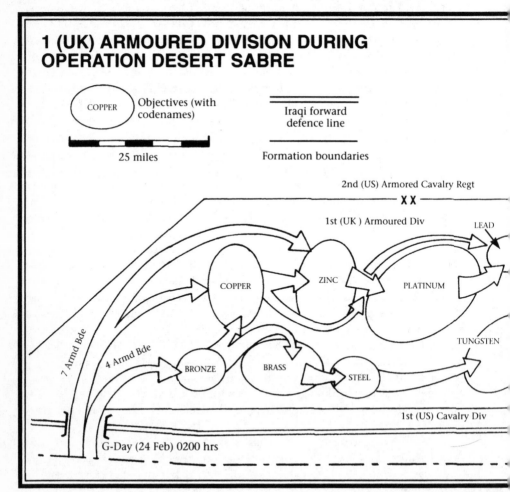

gone on to perform much of its crew training on realistic simulators. Most Iraqis had little or no experience of 'green screen warfare' and many were to die without knowing what had hit them.

Codenamed 'Desert Sabre', the destruction of the Iraqi army in Kuwait commenced at 0400 on 24 February and was completed during the next 100 hours. One British regiment which had already served in Iraq twice since the turn of the century was the 14th/20th King's Hussars. As cavalry, it had played a prominent role in General Maude's capture of Baghdad from the Turks in 1917; and in 1941, equipped with Vickers light tanks, it had again passed through the city to take part in the short but sharp campaign which resulted the elimination of Nazi influence in Iran. Now, in February 1991, it was armed with Challenger MBTs and, under the command of Lieutenant Colonel Michael Vickery, it keenly anticipated entering the city of the caliphs for the third time. The Hussars were the armoured regiment of 4th Armoured Brigade, commanded by Brigadier Christopher Hammerbeck, one of two formations which consti-

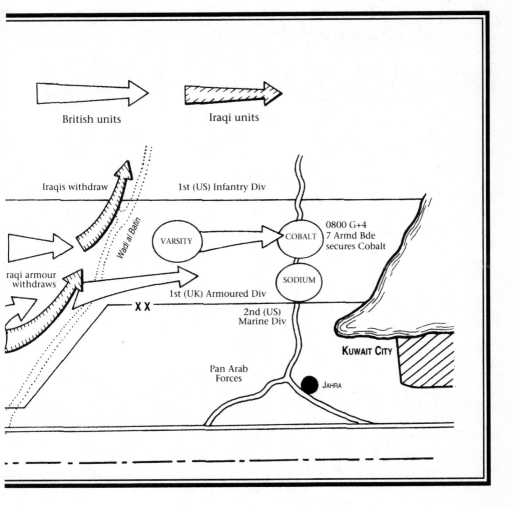

tuted Major General Rupert Smith's British 1st Armoured Division, the other being 7th Armoured Brigade commanded by Brigadier Patrick Cordingley.

The regiment's experience in this, the largest clash of armour since World War II, was common to that of all the participants in the Coalition army, and it is graphically encapsulated in the description of the actions fought by its B Squadron, written shortly after the event by the squadron leader, Major Richard Shirreff.

'The battlegroup mission was to destroy the enemy tactical reserve with the aim of protecting the right flank of US VII Corps. Finally, on 24 February, we were given the order to move. During the morning it became clear that the Americans were moving through the enemy very rapidly. We downed camouflage nets, cleared troop hides – and predictably – while on one hour's notice to move, received the order to move immediately to the transporter pick-up point at 1415. Luckily we were ready to go as we had anticipated such a change in plan from previous experience. No transporters were available so we motored up to the staging area on our own tracks passing long lines of soft-skinned vehicles and other transport heading for the war.

'After a good night in the staging area - and the last sleep for several days - we were given confirmatory orders. The battlegroup mission was to destroy enemy armour and mobility assets within specified boundaries. In outline, once across the start line we were to move to our first objective via a series of way points in standard battlegroup formation - B Squadron forward-right, A Squadron forward-left, D Squadron rear-right, and the Queen's Company Grenadier Guards, travelling in Warrior infantry fighting vehicles rear-right. We were told that it was unlikely that we would meet much enemy en route and that the first major opposition we would encounter would be in Objective Brass. In the event, the only serious opposition encountered was en route and the enemy in Brass offered no resistance whatever!

'After a quick squadron orders group we moved out of the staging area at about 1330. We closed down as we left and moved with all NBC pressurisation switched on and wearing NBC kit; at this stage there was a very real fear of a chemical attack. The squadron log records crossing the Iraqi frontier at 1438 on 25 February. On arrival at the forming up point at about 1730 the squadron deployed into troop columns and replenished all tanks. 7th Armoured Brigade were still moving ahead and their logistic tail was on the move all round us. It was an extraordinary sight: a broad swathe of vehicles of every description moving across the desert. It was pouring with rain, thick and black from the Kuwaiti oil fires and next to us an M109 self-propelled artillery battery and multi-launch rocket systems (MLRS) were firing in support of 7th Armoured Brigade and softening up our own objectives. At this stage it was getting dark and starting to rain heavily. Visibility was very bad.

'At about 1845 we were told that H Hour was confirmed as 1930. We moved off on time, linked up with A Squadron on our left and, after threading our way in squadron column through what we assumed to be the echelons of 7th Armoured Brigade, crossed Phase Line New Jersey (the start line) and went into action. The advance to Objective Brass then began in earnest with B Squadron forward-right squadron of the battlegroup. As we broke free of the clutter of logistic vehicles we deployed into troop line. 1 Troop under Staff Sergeant Fogg was forward-left, the Squadron Leader forward-centre, and 2 Troop under Lieutenant John Dingley, forward-right. 4 Troop provided flank protection on the right in troop column under Lieutenant Edward Gimlette while 3 Troop under Lieutenant Jonathan Hollands was left-rear. Captain Ian Thomas, the squadron's Second in Command, was centre-rear and Captain Henry Joynson with the squadron administrative packet, including the ambulance, the M548 support vehicle and the Fitter Section, brought up the rear. Despite foul weather, poor visibility and darkness, we made fast progress across the desert, thanks to our satellite navigation system and thermal image night sights.

'At 2227 the first contact was made by Lieutenant Gimlette when he reported enemy infantry and APCs on the right flank at a distance of 1200 metres. Initially, it was thought that the contact may have been a reconnaissance party from our own divisional artillery group and we were ordered to hold firm while the situation was clarified. Meanwhile, the Squadron Leader ordered 4 Troop forward to investigate. As they did so they were illuminated by an IR searchlight, suggesting that a T55 was observing them. At the same time, we received confirmation that no friendly forces were in the area of our contact and that we should open fire. The Squadron Leader ordered 2 Troop to open fire and allow 4 Troop to withdraw into the squadron firing line and engage the enemy. The first tank to open fire was the troop leader's, with a HESH engagement against infantry in the open. [HESH = High Explosive Squash Head, an HE round which can also be used against AFVs; the nose of the round pancakes on impact and the subsequent explosion blasts a scab of metal off the interior of the target's armour.] Meanwhile, 1 and 2 Troops were ordered to swing right and take up positions on the left of the squadron firing line. We were now at 90 degrees to the axis of advance, which was due east, engaging the enemy to our south. As troops came into the line and identified targets, tanks engaged the enemy to our front with a mixture of APFSDS [Armour Piercing Fin Stabilised Discarding Sabot], because most tanks had that loaded already, HESH and machine-gun. The engagement continued for about an hour during which approximately eight transport targets were destroyed on the left. On the right, several trench systems were engaged and at least one section of infantry destroyed with a well placed HESH round. In addition, a BTR 60PA and two MTLBs armoured personnel carriers were destroyed. The enemy returned fire, principally with machine-guns. However, Lieutenant Dingley's tank was engaged by two rounds of anti-tank fire and Captain Thomas's tank was illuminated by IR searchlights, although no T55s were subsequently discovered on the position.

'Once it was clear that the enemy were no longer interested in returning fire, Colonel Vickery ordered the Queen's Company up behind B Squadron and told Major Shirreff to lead them onto the position to clear it and take prisoners. Once the Queen's Company had redeployed, the squadron moved forward under illumination from the artillery. We advanced slowly with guns traversing menacingly and headlights on. As we did, we were confronted by large numbers of Iraqis jumping out of trenches waving white flags and holding up their hands in surrender. Captain Joynson and Squadron Sergeant Major Rae initially took their surrender but soon the Queen's Company took over the task and sorted them out in good Footguards fashion. Artillery illumination gave us a clear idea of the position. It was large and covered with a complex of trenches together with dug-in vehicles and APCs. Subsequently we discovered that we had overrun two positions; a transport echelon on the left, possibly from an artillery unit, and what appeared to be a signals unit – both, presumably, supporting the infantry formations dug in to their south manning the border defences.

'At about 0245 (26 February) B Squadron moved back onto the battlegroup axis, linked up with A Squadron and prepared to continue the advance east; the mission remained unchanged. The battlegroup advance was resumed at 0315. B Squadron continued in troop line (two troops up in box formation) but 4 Troop took up position as forward-right, 2 Troop dropping back to rear-right and flank protection because of the ammunition they had expended on the first objective. At 0415 contact was made by Corporal Adesile on the right. He reported a large articulated lorry to his front. This was rapidly engaged and destroyed with one round of APFSDS. From the dramatic fireball that followed, we assumed that this was carrying ammunition. Soon afterwards, we had multiple contacts of tank turrets dug in to our front. Major Shirreff ordered squadron line to be formed and all tanks were quickly in action. We had the bulk of a tank company to our front and soon there were several T55s burning. The ammunition on board them continued to explode as the fires took hold inside. Several enemy tanks returned fire but fortunately no Challengers were hit. While B Squadron was thus engaged, A Squadron and the Queen's Company cleared the position on our left which also included a number of T55s.

'At first light, we had a clearer view of our night's work. Burning tank hulks and the debris of battle littered the desert; disconsolate groups of Iraqis wandered about in a daze looking for someone to accept their surrender. At about 0700 the commander of the Iraqi 52nd Tank Brigade surrendered to Captain Joynson, together with a large number of other Iraqis, presumably his brigade staff. All the T55s destroyed by B Squadron with APFSDS or HESH had been dug in facing south and most had died with their guns traversed right as they faced the sudden, deadly threat from the flank. A few had been destroyed as they attempted to reverse out of their tank scrapes. Seldom can the dangers of digging tanks in as mobile pill boxes have been more graphically illustrated.

'Once the position on the left had been cleared, the battlegroup reorganised prior to continuing the advance to Objective Brass. Colonel Vickery ordered D Squadron to take B Squadron's place as forward-right squadron. B Squadron were tasked as left flank protection for the forthcoming attack, the reason being that we had already been in two actions and, with the main battle still anticipated, there was a danger that we should be short of ammunition. Once A and D Squadrons were clear we moved, turning south at the predesignated waypoint, and halted near a burning BRDM-2 armoured car, waiting for the artillery preparation which would precede the assault on Brass. As we did so, a further group of about forty Iraqis surrendered.

'A and D Squadrons moved at H Hour and we followed. While we waited, 2 Troop engaged three T55s on the right and then spotted, in the same direction, a column of vehicles which included a bridgelayer. Fortunately, the latter was quickly recognised as a Chieftain, although the incident demonstrated the potential danger of clashing with friendly forces in fast-moving desert operations. The column was, in fact, the battlegroup's A1 Echelon.

'When we moved, we swung left to protect the battlegroup's left flank. As we advanced, we contacted a company of eight T55s dug in facing south. Two of them had been destroyed by air attack but the remainder were untouched and, surprisingly, unmanned. We halted, destroyed them, and continued our advance to the limit of exploitation. We went firm on a convenient – and in that featureless country, rare – ridge-line and saw to our front at a range of two to three thousand metres a line of anti-aircraft guns and transport, all obligingly facing south away from us. We engaged these with HESH and scored some good hits. In particular, Trooper Needham, the gunner of Call-sign 32, scored a good first round hit on a transport target at 2700 metres. It was not long before what we now recognised as the Iraqi immediate contact drill – the frantic waving of a white flag – was evident and we stopped firing. Soon, a long column of about 200 Iraqis walked in across the desert, looking about as inspiring as a Greenham Common peace demonstration. While the prisoners were disarmed and sorted out by 4 Troop, Squadron Quartermaster Sergeant Lee and his valiant A1 Echelon, as ever in the right place at the right time, caught up with us. We rebombed with ammunition, filled up with fuel and sorted out our maintenance problems; we also had a much-needed meal. During this action we also captured 150 prisoners from Brass; many were tank crewmen who had dismounted from their tanks and hidden in bunkers while we fought through the position. It later emerged that at least two T55s had fled south as we attacked from the north.

'Once we were administratively balanced again, we received a warning order for the next phase of the operation. B Squadron was to regroup under the command of 3rd Royal Regiment of Fusiliers, who were the right-forward battlegroup in the brigade attack on Objective Tungsten. We were given locations for the large numbers of enemy troops expected there and told to get going. We moved immediately and, after travelling ten kilometres across the desert, linked up with the Fusiliers by 2100. Our task was to advance as left-forward squadron ahead of C Company. The 1st Royal Scots battlegroup was on our left, where MLRS batteries were roaring away and, to the steady crump of artillery landing on the enemy positions, we moved off on the approach march at 2230. The reconnaissance group led, followed by B and D Squadrons in columns, with the infantry companies bringing up the rear in their Warriors. The first contact came when the recce group encountered enemy dug in around an oil installation to our front. A few bursts of machine-gun fire soon persuaded them to surrender. Thereafter, B Squadron crossed an oil pipeline and went firm in a bridgehead about 3000 metres beyond while D Squadron also crossed and shook out to our right. While this was happening, we detected an enemy tank company to our front and engaged and destroyed at least five T55s using thermal image gunsights. We could see, and hear, D Squadron on our right doing the same. We moved on in troop line up to the limit of exploitation in what was almost a classic squadron advance to contact. We moved a bound, made contact, formed line and engaged the enemy to our front. Once they were destroyed, we moved on again until more enemy were contacted, whereupon the same process was repeated. Having arrived at the limit of penetration, we formed a hasty counter-penetration line as left-for-ward squadron. This was completed by 0500 on 27 February.

'Although we had destroyed a considerable quantity of enemy armour on Tungsten, we were faced with almost no opposition. Enemy encountered were two tank companies, one of which was destroyed by B and D Squadrons together, a 152mm howitzer battery, a mechanised infantry company and a logistic installation. Several vehicles were engaged while withdrawing and we believe the bulk of the AFVs engaged were empty with crews hiding until the storm of fire had passed and they could find someone to accept their surrender. During the attack on Tungsten we had one tank immobilised by, it is thought, friendly fire from the Royal Scots on our left. Callsign 30, commanded by Lieutenant Hollands, fell into a tank scrape and while trying to extricate himself was hit in the rear by a 30mm AP round which penetrated the thin rear armour and damaged the final drive. Callsigns 10 and OB also broke down, evidence of the considerable mileage we had done.

'At 0800 we received orders to regroup under the command of 14th/20th Hussars battlegroup. We moved, towing OB, and rejoined the battlegroup for some much needed maintenance. At this stage, we had most of the Fitter Section spread across the desert recovering various tanks and we were also badly in need of replenishment before we could continue the action. By 1330 we were off, advancing north-west towards the Kuwaiti border. No clear objective was given, but the aim was to continue the pursuit of a defeated enemy.

'We shook out rapidly and soon encountered enemy positions, all empty, but with considerable numbers of vehicles left intact. As we advanced, we passed through between 800 and 1000 of the enemy spread out across the battlegroup's axis of advance. We were unable to stop to take their surrender, so we merely stood up in our turrets and pointed to the rear. They quickly got the message and set off for captivity. As we moved forward we engaged several targets to our front. We then contacted a large column of vehicles. While the problem of identification was being sorted out, B Squadron formed line and prepared to engage what we were convinced was a ZSU-23-4 self-propelled anti-aircraft mounting at a range of about 2500 metres. Subsequently we realised that we had been on the point of engaging friendly forces. We were then ordered to go firm and await orders. We were given a 'no move before first light,' so rapidly settled down for the first sleep for three nights.

'During the night we received radio orders that we were to advance east to cut off the Iraqi retreat north of Kuwait City, moving early next morning. We were also rejoined by all our missing tanks, except for Callsign 10. We were therefore well balanced, refreshed and ready to continue the battle. Sadly, the Iraqis were not of the same spirit and by now it was known that a ceasefire was imminent.

'We moved off as a battlegroup early on 28 February in column. It was a hazy morning and visibility was down to less than 1000 metres. We were slowed down crossing the Wadi al-Batin and a series of quarries on the eastern side. However, we kept up a good pace and were soon into Kuwait. Twenty kilometres east of the wadi we encountered numerous enemy positions and abandoned armour. At 0800 we were told that a ceasefire had been announced and at 0830 we were halted. B Squadron's war was over.'

Despite having destroyed several times its own weight in enemy armour and removed numerous units from the Iraqi order of battle, B Squadron's only personnel casualty was a broken leg sustained in a fall.

Elsewhere, the US XVIII Airborne Corps' strike into the Euphrates valley, combining as it did airmobile operations with rapid ground advances, caught the Iraqis completely off balance and, as Schwarzkopf had intended, left Saddam's forces in Kuwait completely isolated save for one narrow axis of withdrawal through Basra. Attempts by several Republican Guard divisions to intervene against US VII Corps were very roughly handled by the American armour and the British 7th Armoured Brigade, leaving the impression that they were very ordinary soldiers indeed. On the southern sector the US Marines and Arab national contingents experienced no difficulty in breaking through and were soon advancing on Kuwait City through the broken wreck of the Iraqi army. Those Iraqis that could scrambled aboard whatever transport was to hand, including thousands of civilian vehicles, and headed north towards Basra; caught in the mother of hysterical traffic jams, they were pounded into a tangle of twisted metal by the Coalition air forces.

By the time the ceasefire came into force, 42 Iraqi divisions had been destroyed as fighting formations. Estimates of Iraqi killed varied between 25,000 and 100,000; the precise figure will never be known, but may be in the region of 60,000. About 80,000 prisoners were taken and a similar if not larger number may have simply deserted. 3700 tanks, 2400 APCs and 3000 artillery weapons were destroyed or captured. The gloomiest of Western commentators had predicted that the Coalition army would sustain between 10,000 and 50,000 casualties in the land battle. In the event, the total came to less than 500, including 150 killed; of the latter, 28 died in a Scud attack on the Dhahran air base and a high proportion of the remainder were killed in 'friendly fire' battlefield accidents.

Although Desert Sabre fulfilled the UN mandate for the liberation of Kuwait, many believe that President George Bush's imposition of a ceasefire was premature. Perhaps as many as 20,000 Iraqis with approximately 700 tanks and 1400 APCs managed to escape the débâcle and were immediately used by Saddam to put down rebellions by Kurds in the north of the country and Shi'ites in the south, using the utmost ferocity. These ruthless measures ensured that he remained in power and the very fact of his survival enabled him to convince his countrymen that he had won a victory.

This sour note apart, Desert Sabre had demonstrated a classic application of the blitzkrieg technique in a desert environment. Seventy-five years had passed since the first primitive tanks had ground their way slowly towards the enemy's trenches yet, if the shades of Fuller and the other tank pioneers were watching, they would have seen most of their own ideas more fully translated into action than in any previous campaign, short though the land battle for Kuwait was. The Coalition army gathered together many of the strands in the history of armoured warfare, and it was just the sort of army they had dreamed of, in which everyone, tanks, infantry, artillery, engineers and even medical services, advanced together on tracks and under armour, using the indirect approach to achieve the strategic paralysis of the enemy and the destruction of his army. Above all, the tank had been invented to save lives and, as can be seen from the Coalition casualty figures quoted above, that is what it had done.

Select Bibliography

Anon, *The Story of 79th Armoured Division*, privately published by the Division, July 1945

Beddington, Major General W. R., *The Queen's Bays 1929–1945*, Wykeham Press, 1954

Carius, Otto, *Tiger im Schlamm*, (ISBN 3-921-655-42-0)

Carver, Michael, *El Alamein*, Batsford, 1962

Cooper, Bryan, *The Ironclads of Cambrai*, Souvenir Press, 1967; *Tank Battles of World War I*, Ian Allan, 1974

Crow, Duncan, *US Armor-Cavalry – A Short History 1917–1967*, Profile, 1973

Dawnay, Brigadier D., et al, *The 10th Royal Hussars in the Second World War 1939–1945*, Gale & Polden, 1948

Duncan, Nigel, *79th Armoured Division – Hobo's Funnies*, Profile, 1972

Dunstan, Simon, *Vietnam Tracks – Armor in Battle 1945-75*, Osprey, 1982

Ellis, Major L. F., et al, *Victory in the West Vols 1 & 2*, HMSO, 1962 and 1968

Elstob, Peter, *Bastogne – The Road Block*, Macdonald, 1968

Erickson, John, *The Road to Stalingrad*, Weidenfeld & Nicolson, 1975; *The Road to Berlin*, Weidenfeld & Nicolson, 1983

Eshel, David, *Chariots of the Desert – The Story of the Israeli Armoured Corps*, Brassey's, 1989

Fletcher, David, *War Cars – British Armoured Cars in the First World War*, HMSO, 1987

Foley, John, *The Boilerplate War*, Frederick Muller 1963

Forty, George, *United States Tanks of World War II*, Blandford, 1983

— *German Tanks of World War Two*, Blandford 1988

— *A Pictorial History of the Royal Tank Regiment*, Guild, 1989

Fraschha, Gunther, *Mit Schwerten und Brillanten*, (ISBN 3-8004-1176-8)

Gaujac, Colonel Paul, Ed, *Sedan 1940*, Service Historique de l'Armée de Terre

German Report Series, US Department of the Army: Pamphlet No. 20-230, *Russian Combat Methods in World War II*, 1950; Pamphlet No. 20-233, *German Defense Tactics Against Russian Breakthroughs*, 1951; Pamphlet No. 20-234, *Operations of Encircled Forces – German Experiences in Russia*, 1952;

Guderian, Major General Heinz, *Achtung – Panzer!*, Arms & Armour, 1992

Herzog, Chaim, *The Arab-Israeli Wars*, Arms & Armour, 1982

Hopkins, Major General R. N. L., *Australian Armour - A History of the Royal Australian Armoured Corps 1927–1972*, Australian War Memorial 1978

Horne, Alistair, *To Lose a Battle – France 1940*, Penguin, 1979

Jukes, Geoffrey, *Kursk – The Clash of Armour*, Macdonald, 1968

Klein, Egon and Kuhn, Volkmar, *Tiger – Die Geschichte einer Legendären Waffe 1942–45*, Motorbuch Verlag, 1976

Larionov, V., et al, *Decisive Battles of the Soviet Army*, Progress Publishers, Moscow, 1984

Liddell Hart, Captain B. H., *The Tanks – The History of the Royal Tank Regiment*, Vols 1 & 2, Cassell, 1959

Lucas, James, *War on the Eastern Front 1941–1945 – The German Soldier in Russia*, Jane's 1979

Lucas Phillips, C. E., *Alamein*, William Heinemann, 1962

Lucke, Christian von, *Die Geschichte des Panzer-Regiment 2*, Boss-Druck und Verlag, 1953.

MacDonald, Charles B., *The Battle of the Bulge*, Weidenfeld & Nicolson, 1984

McAulay, Lex, *The Battle of Coral*, Arrow, 1990

McCarthy, Chris, *The Somme – The Day by Day Account*, Arms & Armour, 1993

Macksey, Kenneth, *The Guinness Book of Tank Facts and Feats*, Guinness Superlatives, 1972

Orde, Roden, *History of the Second Household Cavalry Regiment*, Gale & Polden, 1953

Orgill, Douglas, *The Gothic Line*, William Heinemann, 1967

Perrett, Bryan, *The Churchill*, Ian Allan, 1974

— *A History of Blitzkrieg*, Robert Hale, 1983

— *Knights of the Black Cross – Hitler's Panzerwaffe and its Leaders*, Robert Hale, 1986

— *Soviet Armour Since 1945*, Blandford, 1987

— *Desert Warfare*, Patrick Stephens, 1988

— *Canopy of War – A History of Jungle Warfare*, Patrick Stephens, 1990

— *Tank Warfare*, Arms & Armour, 1990

— *Tank Tracks to Rangoon*, Robert Hale, 1992

— with Anthony Lord, *The Czar's British Squadron*, William Kimber, 1981

Pimlott, John and Badsey, Stephen, Ed, *The Gulf War Assessed*, Arms & Armour, 1992

Pitt, Barrie, *1918 – The Last Act*, Cassell, 1962

Pitt, Lieutenant Colonel P. W., *Royal Wilts – A History of the Royal Wiltshire Yeomanry 1920–1946*

Playfair, Major General I. S. O., et al, *The Mediterranean and Middle East, Vol 4*, HMSO, 1966

Ryan, Cornelius, *A Bridge Too Far*, Hamish Hamilton, 1974

Schaufler, Hans, *So Lebten und so Starben Sie – Das Buch vom Panzer-Regiment 35*, Kameradschaft ehem. Panzer-Regiment 35

Schneider, Wolfgang and Strasheim, Rainer, *Deutsche Kampfwagen in 1. Weltkrieg*, Podzun-Pallas-Verlag, 1988

Smith, Robert Ross, *Triumph in the Philippines*, Center of Military History, US Army, 1963

Starry, General Donn A., *Armoured Combat in Vietnam*, Blandford, 1981

Terraine, John, *The Smoke and the Fire - Myths and Anti-Myths of War 1861–1945*, Sidgwick & Jackson, 1980

— *White Heat – The New Warfare 1914–18*, Sidgwick & Jackson, 1982

Teveth, Shabtai, *The Tanks of Tammuz*, Weidenfeld & Nicolson, 1969

Toland, John, *Battle – The Story of the Bulge*, Severn House, 1977

Williams, John, *France 1940*, Macdonald, 1969

Zhukov, Marshal Georgi, et al, *Battles Hitler Lost*, Richardson & Steirman, 1986.

Periodicals

The Hawk, Regimental Journal of the 14th/20th King's Hussars, 1991

Das Schwarze Barett, Journal of the German Armoured Corps, No. 9

Index